Social and Economic Statistics for Africa

Their Sources, Collection, Uses and Reliability

G.M.K. Kpedekpo and P.L. Arya

Routledge
Taylor & Francis Group
ROUTLEDGE REVIVALS

First published in 1981 by George Allen & Unwin Ltd

This edition first published in 2024 by Routledge
4 Park Square, Milton Park, Abingdon, Oxon, OX14 4RN
and by Routledge
605 Third Avenue, New York, NY 10158.

Routledge is an imprint of the Taylor & Francis Group, an informa business

ISBN 13: 978-1-032-74701-9 (hbk)
ISBN 13: 978-1-003-47053-3 (ebk)
ISBN 13: 978-1-032-74720-0 (pbk)
Book DOI 10.4324/9781003470533

Social and Economic Statistics for Africa

Their Sources, Collection, Uses and Reliability

G. M. K. KPEDEKPO
Professor of Economics, Ahmadu Bello University

and

P. L. ARYA
Senior Lecturer, Ahmadu Bello University

London
GEORGE ALLEN & UNWIN
Boston Sydney

First published in 1981

GEORGE ALLEN & UNWIN LTD
40 Museum Street, London WC1A 1LU

British Library Cataloguing in Publication Data

Kpedekpo, G. M. K.
Social and economic statistics for Africa
1. Africa - Statistics
I. Arya, P. L.
316 HA1955

ISBN 0-04-310011-2

Set in 10 on 11 point Times by Katerprint Ltd and printed in Great Britain by Billing and Sons Ltd., Guildford, London and Worcester.

Contents

Preface

The phenomenal growth in the teaching of statistics in African universities to social science students is not limited to the increase in the number of universities now offering statistics courses to their students, but also includes the expansion and widening of the scope and range of topics that have been included in the syllabuses of these departments.

These developments have been much more striking in the teaching of statistical theory than in applied statistics. Undoubtedly, one of the strongest practical reasons for the heavy loading of mathematical theory in statistical courses is the difficulty of examining anything else. The second reason is that there are numerous books on statistical theory that both lecturers and students can have access to.

However, students who have gone through these courses are often 'disillusioned with our introductory courses' not only in Africa but also in the United Kingdom, United States and elsewhere. They are 'turned off by statistics' and 'less numerate at the end of the course than at the beginning'. This happens whatever their main subject (biology, economics, management, medicine, sociology, psychology, etc.) and largely irrespective of whether the course is run by a subject matter specialist or by someone from a statistics department, and of how it is organised. A particular complaint is that most of the courses seem 'irrelevant to what we were doing'.

There are, no doubt, several neglected branches of statistics or topics that lend themselves to an almost completely non-abstract approach and that have practical relevance to the student. The first neglected topic is information on the types of source material available on social and economic statistics in Africa (their errors and limitations). The second neglected topic is that of interpreting published figures, say crime figures, using data from the annual abstract of statistics or the statistical digest or police records. Of course, new kinds of statistics courses or topics will require new books and possibly even new professors!

The primary purpose of this book, therefore, is to introduce undergraduates in the first and second years of their introductory statistics courses to source material that is used extensively by practising statisticians in government and research organisations. A secondary objective is to fill an important gap in the provision of teaching material in this area. The emphasis is on source material on African social and economic statistics. Stress is laid on the kinds of sources available in Africa to enable the student to search for these in the appropriate ministries or government departments, rather than on providing the details on each topic for each African country, a task that is clearly a daunting one. Particular stress is laid on the role that the introduction of Western concepts and definitions plays in the collection, analysis and interpretation of African statistics and on the problems that these give rise to.

The book is organised in thirteen chapters. Chapter 1 deals with the general nature of social and economic statistics in Africa: their sources, collection, uses and limitations. Chapters 2 to 12 treat specific topics in social and economic statistics: population, crime, health, housing, labour, social security systems, agriculture, external trade, finance, national accounts. The treatment of each topic goes into detail on the nature and scope of the statistics, sources, uses and limitations. The final chapter discusses the concept of 'indicators' and attempts to pull together the various ratios and rates that can be used for measuring development.

This book will, it is hoped, make a useful contribution to undergraduate courses in statistics in African universities, especially when used in conjunction with such books as: *Introductory Statistics* (Biji Afonja), *Modern Elementary Statistics* (J. E. Freund), *Statistics – Schaum's Outline Series* (M. S. Spiegel) and *Introductory Statistics for Business and Economics* (Wonnacott and Wonnacott), which concentrate more exclusively on theoretical aspects of statistics. This book's brevity and relative simplicity can also make it a useful reference work for any interested reader.

Economics Department, G. M. K. KPEDEKPO
Ahmadu Bello University, P. L. ARYA
Zaria, Nigeria

Acknowledgements

The completion of this book would have been impossible without the assistance of many institutions and individuals. We received much help from the university departments where the authors were staff members. We wish therefore to express our gratitude to our colleagues at the Institute of Statistics and Applied Economics, Makerere University; the Department of Economics, Ahmadu Bello University; the Institute of Statistical, Social and Economic Research, University of Ghana; the School of Economics, Delhi, India; and the University of Calabar, Nigeria, where several discussions took place on the restructuring of statistics syllabuses and the teaching of statistics, and where interest and encouragement were expressed.

We are also grateful to anyone who has worked in this field and acknowledge our debt of gratitude to all those persons whose results we have drawn upon during the writing of this book; in particular we are grateful to Heinemann Ltd for permission to quote extensively from *Development Economics in Action: A Study of Economics Policies in Ghana* by Tony Killick (1978) and Princeton University Press for permission to reproduce sections of *The Labour Force in Economic Development: A Comparison of International Census Data 1946–1966* by J. D. Durand (1975).

Another debt of gratitude is owed to Margaret Kpedekpo for her able and competent secretarial assistance freely offered both in typing and in checking the manuscript. She has at various times also acted as research assistant. In fact, the book would never have seen the light of the day but for her assistance in correcting the whole typescript before it was sent to the publisher. Any errors of omission and shortcomings of the book are of course our responsibility alone. Finally, the views and opinions expressed in the book are our own and do not in any way reflect the views of any organisation, past or present, with whom we have been associated.

Department of Economics,
Ahmadu Bello University,
Zaria, Nigeria

G. M. K. KPEDEKPO
P. L. ARYA

CHAPTER 1

Nature, Scope, Uses and Limitations of Social and Economic Statistics

The origin of the word 'statistics' lies in the description of a state as a system or political entity. This usage persists in the field of government statistics, where large numbers of statisticians are employed to compile records and summaries of those records that describe the activities of the state. However, the term 'statistics' can also be used in two forms: singular and plural. In the singular form it denotes the techniques of statistical analysis. These are measures of location, variation, probability, correlation and regression, analysis of variance, and so on. In the plural form it refers to statistical data. Data are a set of figures: for instance, data on production, population, employment, prices, wages, interest rates, exports and imports.

For a proper analysis of the social and economic conditions prevailing in any country at a particular period of time, knowledge of both aspects of statistics, techniques of statistical analysis and a proper assessment of the quality of the data, is essential. If a set of data is given to a layman, he may not be able to derive conclusions; but the social scientist, on the other hand, who is equipped with the techniques of analysis, should be able to derive certain concrete results from them. However, the validity of these results will depend on the coverage and the reliability of the data. A social scientist, who knows only the techniques of analysis, but does not take the trouble to learn something of the sources and reliability of data, may come out with inappropriate conclusions or wrong judgements. This is precisely the reason why the syllabuses of all the universities emphasise the techniques of analysis and the sources and methods of collection of data.

This introductory chapter outlines broadly the sources, methods and uses of the data relating to African countries, while separate chapters are devoted to the detailed analysis of topics in social and economic statistics.

SOURCES OF AFRICAN STATISTICS

There are two main sources from which data on social and economic statistics can be obtained in Africa: published and unpublished sources. These sources can be further classified into two groups: national and international sources. First, we consider published national sources.

Published National Sources

The national sources of data relate to the data published by the statistical offices of different countries, by research institutions and by individual researchers. The main sources of such published data are as follows:

(1) *Statistical Abstracts, Bulletins and Reports Issued by Government Departments, Especially Their Statistical Units*

The central statistical organisation of African countries has been assigned the responsibility of publishing data relating to the particular country's economy. These organisations periodically bring out publications for this purpose. The frequency of publication ranges from one month to one year. However, the coverage of the various aspects of the economy depends on the availability of resources (both human and financial). The bulk of most of these publications is generally made up of tables with explanatory notes derived from data collected by the department itself or by some other agencies. They can properly be described as data source publications.

In describing these data sources, two terms are usually employed: 'primary sources' and 'secondary sources'. The former is used in situations where data are collected and published by the same agency (primary source), and the latter is used when the publishing agency is not the one that collected the data (secondary source). These publications cover a very wide spectrum of topics about social and economic conditions, ranging over population, crime, exports and imports, national accounts, housing, health, and so on. The most exhaustive single source of national published data is the statistical abstract or the statistical digest, published annually by many countries.

(2) *Miscellaneous Reports of Government and Non-Government Agencies*

These are not reports specifically made up of tables. The tables that feature in such reports are normally incidental and used mainly to summarise the activities of the agency over the period covered by the reports: for example, produce-marketing boards, insurance corporations and universities.

The central banks of these countries, for example, publish data relating to the money and finance, production, prices, balance of payments, foreign exchange and general aspects of the economy. Apart from these government publications there are publications issued by the private sector. The banks and other financial institutions are obliged by law to publish their statement of accounts.

(3) *Research Reports and Publications in Learned Journals*
These are usually devoted to reports of specific research investigations. Research institutions carry out independent surveys and publish the results thus obtained.

Individual researchers also carry out independent surveys on particular fields of interest. The results of their research are published in either academic journals or books. The coverage of these private sources of data, however, is limited to a particular aspect of the economy; nevertheless, they provide a useful cross-check when compared with the official data, when available, or fill up the important gaps in the data.

(4) *Daily Newspapers, Magazines and Miscellaneous Periodicals*
It is difficult to make any generalisation about the authenticity of data from these sources. Some fairly reliable data, especially of a social science character, do occasionally appear on the pages of some of them, but sometimes data can be manipulated to make their stories sensational. As a general rule, a great deal of caution is required in using data based on these sources.

Published International Sources

The government statistical documents are rarely sufficient, but much additional information can usually be obtained from the documents published by the various international organisations. These international sources of data consist of the publications of the international institutions: the International Bank for Reconstruction and Development (IBRD, the World Bank), the International Monetary Fund (IMF), and other United Nations (UN) agencies including the Economic Commission for Africa (ECA).

The characteristics of the statistics published by the three most important organisations (see, for example, Bondestam, 1973) – the UN, the ECA and the IBRD – are:

- UN: The information given is a summary of statistics by the various governments, which is collected yearly.

- ECA: Statistics are frequently collected from various African governments that are members of the ECA and analysed by ECA personnel.
- IBRD: Statistics are mostly collected over a short time period by experts in the respective country from government organisations. Due to the particular method of collection, the IBRD reports are usually unreliable.

In general, the international agencies cull data from the statistical offices and central banks of different countries and publish these periodically. Sometimes, however, they do carry out their own independent surveys. The uses of their publications are numerous. For example, one can obtain a comparative picture of major aspects of the social and economic situation of many countries in a single publication. The independent surveys carried out by the international agencies are quite useful, since they provide an important cross-check on the data issued by the national statistical agency, fill in the existing gaps in the data and provide international comparisons of the activity surveyed.

The problem that often arises in data collection and analysis on any topic in statistics is how to reconcile the needs for international comparability with the specific domestic needs of individual countries. The value of this kind of comparability for analytical purposes can hardly be exaggerated. In the field of social and economic statistics, for example, an attempt is always made to find relationships between certain factors by examining the experience of other similar countries, despite the fact that such comparisons are fraught with technical difficulties, particularly of interpretation.

The frequency of publication of the international statistical abstracts also varies from one month to one year, as in the case of the official publications of the different countries. The most exhaustive monthly source of international data is the monthly bulletin of statistics, and the most comprehensive source of information is the statistical year book, which is designed to serve as a convenient coherent summary of all international, economic and social statistics.

Unpublished Sources

Data in their original form exist in the files and books and on various forms of many government and non-government departments and corporations. Such data, although not published, can be used with profit by researchers to cross-check other information available in published form. But the task of obtaining such information in Africa can be tremendous and time consuming, partly because the storage of such unpublished data leaves much to be desired, and partly because of confidentiality regulations concerning access to such government information. The main difference between these two sources is that published data are more readily accessible than unpublished ones.

DATA COLLECTION METHODS

The concepts, definitions and methods of collecting African statistics are not very different from those used in developed countries because of the Western colonial heritage, although these should of necessity vary since conditions are so different. This is one reason why statistics in Africa are less reliable than those found in developed countries. But before we make any more critical comments on the end product of these methods of data collection – the statistics – let us deal briefly with the methods of data collection in general.

Data are collected by either sample or census methods. In the former, a particular area is taken as representative of other areas, and the data relating to it are taken as a replica of the rest of the population. The census method is more exhaustive and covers the entire economy. The census of production and the census of population are examples of the latter method, while the savings surveys and consumption surveys are examples of the former method. The difference in the two methods is only the coverage. The census covers the entire economy under study, while the sample method restricts itself to a particular selected area and assumes it to be the representative of the rest of the country. The results obtained by the sample and the census methods differ. A question arises: why not adopt the census method, which covers the entire economy, and obtain a better view of all aspects of the economy? The answer is quite simple. Since a census covers the entire country, it requires a large number of people to carry out the survey. In cases where a census method is adopted – for example, a population census or an agricultural census – a large number of personnel is required to carry out the data collection; thus, apart from the regular staff of the census bureau, teachers and other government officials are deployed to undertake enumeration duties. African countries already lack skilled manpower or qualified personnel, and it is not always possible to apply the census method, which requires a large educated manpower to carry out surveys. Moreover, such surveys are expensive and time consuming. The sample method is comparatively quicker and less expensive and requires less labour force. The big questions are how to determine what level of costs is justifiable in the production and collection of statistics, and then how to allocate those resources among the many options.

The choice of the sample or census technique also rests on the problem at hand. The census method is adopted where the large regional variations of the object under study are expected, while the sample method is applied where the expected variations are less. Further, the census method can be applied in the case of production of an organised industrial sector, where the industries are required to fill up returns of their output and earnings regularly. In this case it is convenient to apply census methods.

Political factors also determine whether we adopt a sample or census method. For instance, population data are quite sensitive to politics. Every state in the country would prefer that the data of population relating to its regional boundary should not underestimate the population. On the basis of population, it can claim several advantages: increased representation of the region in Parliament, increased central financial assistance, and so on. The sampling technique would not satisfy anybody in this case. Further, the adoption of the sample and the census techniques also depends on which method is followed by other countries in the study of a particular problem.

However, census and sample surveys are not necessarily alternative methods of data collection. They are often complementary and draw on each other in many ways. A sample survey conducted in conjunction with a census can be used in collecting additional data on items that need special investigation techniques or in more probing than can be afforded in a large scale census. Intercensal sample surveys help in making census data up to date. Postcensus sample surveys are used to measure the accuracy of census results. A census often provides the best sampling frame for designing efficient sample surveys and is used in improving estimation procedures.

The data relating to most of the aspects of the economy, published by the central banks and central statistical organisations of African countries, do not require any complicated system for their collection. The suppliers of data are obliged by law to furnish these data to the government for regulation purposes. The scope of these data and their reliability depend upon how many activities are covered by regulation and how effectively they are carried out.

Wherever a survey is required, it should be planned carefully to minimise cost of collection, time and manpower requirements. Planning involves fixing the objectives of the survey, defining the coverage or the population to be studied, deciding the methods of data collection (interviewing method, mailed questionnaires, direct observations, etc.), and choosing the sample design and the questionnaire design. The execution aspect of planning involves the selection and training of interviewers and supervisors, pilot surveys, field organisation, data-processing and writing of the report. The guiding principles in planning and executing a survey must be utility and usability of the data collected. The criterion of utility or usefulness of data will depend on the objectives of the survey and on whether the items of data and universe meet these objectives. The criterion of usability will depend on the objectivity, reliability and timeliness of the data and their presentation. It is important to provide adequate funds, time and research capacity in budgeting for a survey in order to meet the needs of flexible objectives and unforeseen problems that may be encountered.

The methods of collecting statistics described in the previous paragraph are simple, direct and well tried, and there is no reason to doubt that, in future, basic socioeconomic data in Africa and many developing countries will be obtained in these ways. At present, however, there are special circumstances in most countries of Africa that have made the study and the use of alternative procedures necessary and advantageous. Among these special difficulties are the illiteracy of the majority of the population and vagueness about time periods, particularly ages, the difficulty (both physical and mental) of communications, and cultural factors such as the belief that misfortune will follow the disclosure of certain facts. There is also resistance to questioning and falsification of results for economic and political reasons. Although there is no evidence that these are more prevalent in Africa than in other places, awkward problems arise from some of these factors, leading to the provision of special techniques of estimation for dealing with the problem.

METHODS OF ANALYSIS

Analysis of social and economic phenomena can draw on various statistical techniques, but generally extensive use is made of the concepts of ratios and rates. A ratio is a general term given to any numerator – denominator relationship between two numbers; for example, if there were 1,000 students registered within a social science faculty in a university in the 1976 – 7 academic year, and 100 lecturers were also employed within the faculty in the same academic year, the number of students divided by the number of lecturers (a ratio) would give the student – lecturer ratio within the faculty. In this example it would be ten students per lecturer.

Rates, on the other hand, can be regarded as special cases of ratios. They usually measure the likelihood of occurrence of a phenomenon within a given population defined by the spatial and temporal characteristics. Crude rates are obtained by dividing the number of events of a specified type, say crime, occurring within an interval, usually one year (temporal characteristics) by the size of the population within which the events occur. For many purposes, the mid-interval population is taken as the best estimate; for example, the crude crime rate in the city of Accra, Ghana, in, say, 1970 can be calculated as the ratio of crimes committed and detected during a twelve-month period to the population in Accra within the same twelve-month period.

Specific rates are analogous to crude rates but are computed on the basis of events in or to particular subpopulations; for example, age-specific crime rates relate the number of crimes in, say, the 15 – 19 age group to the population in the 15 – 19 age group within the population. These concepts are simple and the calculations easy, but caution is needed in their interpretation. Ratios and rates, although useful in themselves, are also the starting point for more advanced analysis.

USES OF SOCIAL AND ECONOMIC DATA IN PLANNING

Although statistics cannot by themselves be used to determine the social and economic aims that a society should pursue, the construction and implementation of any plan require an immense amount of co-ordinated numerical data drawn up within a coherent framework on population, health, crime, housing, employment, agriculture, education, manpower, and so on. But as we have mentioned, there are gaps in the data that the planner is expected to use, and the available statistics are frequently unreliable and defective. Basic data on population, household formation, patterns of income and consumption, crime, houses and education, and other relevant data, are at best suspect and very often non-existent. But development-planning in the conventional sense is vulnerable to data inadequacies and limitations. The situation is such that the danger of compounding errors is greater when inaccuracies are incorporated in a plan, which by its very nature is required to be internally consistent. In addition, development-planning, where statistics constitute an important part, has chosen a Western-minded course, which again is seldom applicable to African conditions and problems. Accordingly, much of the statistics can never be used effectively. This lack of relevance has been mentioned in connection with the collection of statistics in the field. Thus, the limitations imposed by lack of relevance and paucity of data in Africa, when one begins to consider the uses of social and economic data, are enormous. Nevertheless, we must concede that, whatever statistics are available, they both mould and reflect our perceptions of reality of a given social and economic situation. What, then, are the kinds of social and economic information that one requires, and how can they be used by policy-makers?

Information or data on population as a basis for policy formulation and decision-making should attempt to throw light on the number of people in the country, its geographical distribution and its social and economic characteristics, such as marital status, employment and education. It should also provide information on current levels and trends of fertility and mortality, current population growth, internal migration and urbanisation. Data on crime should identify the characteristics of criminals: that is, their social and economic characteristics – age, sex, employment status and previous history of crime. The data can be used in formulating measures to combat crime in our societies.

Measures of the utilisation and non-utilisation of labour are also of great concern to policy-makers; data on employment and unemployment as well as underemployment are therefore necessary for formulating policies on the effective use of human resources in any country. Housing, as we know, is a basic human need for protection from the weather. The statistics on housing furnish a systematic account of changes in the supply of housing available to the population as well as the natural, economic and social factors that have a bearing on the supply and demand for housing.

In dealing with the health problems of the population, information about hospital beds and their occupancy, the social and medical characteristics of individual patients occupying those beds, the patterns of disease within the population, what age, sex and social groups predominate the particular disease groups, and the use of resources (staff, diagnostic facilities, etc.) are also required. Above all, in the field of health economics and overall planning, there is need to measure resources and relate output to the use of resources. In Africa a large proportion of the population is engaged in agricultural activities. The ability to support the population of a nation largely depends on the success of the agricultural sector. It is an important economic sector, and statistics on agriculture play a dominating role in estimating future production, income, consumption of intermediate and final goods, and employment opportunities.

Finally, national accounts present the records of economic transactions of an economy in a systematic manner and show the relationship between the various components of the economy. The main contribution of statisticians to the economic planning process, therefore, is to construct increasingly detailed systems of national accounts embodying information on input and output, the distribution of income, flow of funds, balance sheets, and so on. These systems provide a data base for increasingly sophisticated economic models, which provide a means of formulating and reviewing policies. It is to the usefulness of models in socioeconomic planning that we next turn our attention.

A model can be defined as a representation of certain aspects of reality designed to highlight the behaviour of some real world phenomena. Most research and much policy formulation are based on single issue models, but socioeconomic systems are complex. For many social and economic planning models, the intention is to obtain a coherent and consistent set of policy interventions. First, a quantitative model can be used to analyse the impact of a specific intervention – policy – by conducting hypothetical experiments representing different policy choices, comparing the outcomes and identifying those policies that are most desirable or least desirable according to some defined criteria.

Models can also be used to ensure consistency in planning by checking intersectoral consistency, ensuring that income, production and human resource targets are compatible, and relating government revenue to government expenditure. Models can be used to investigate the impact not only of individual policies but also of clusters of policies that together form development strategies, by investigating the simultaneous impact of complementary policies, identifying not only the effects of each individually but also their interaction effects. The development and use of models will thus in principle lead to improved policy selection.

A second important role that models play in planning is to identify data lacunae. Given that a problem is well defined, a list of data needed for building the model can be developed. This list can often be reduced once the model has been built and the crucial relationships identified, but it will frequently lead to the identification of gaps in the data needed for studying a particular problem. Thus, the formulation of a model can help the statistical services and related bodies to identify gaps in their data collection and, equally important, problems in the consistency or quality of the data that they collect. In addition, sensibility analysis with a model can lead to a ranking of different data in terms of importance.

There are, however, general and specific reservations about the use of models. These range from the size and complexity of specific models to the fact that in most developing countries the data base is rather inadequate. In many countries of Africa, for example, the empirical base consists of a very restricted range of inaccurate information, and this means that caution is needed in interpreting model outcomes. Full exploitation of existing data, together with judicious development of new data sources, can considerably reduce the problem.

LIMITATIONS TO THE UTILISATION OF AFRICAN STATISTICS

The main sources of data that are used by governments and by economic researchers are official statistics and statistics collected through surveys. Official statistics are incidental to the process of administration and are collected continuously by various government departments and institutions. They do not cover all the areas on which information is needed for guiding policies, and even in the areas for which they are available, the data tend to be incomplete, inconsistent (because uniform concepts and definitions are not specified), biased (because they are affected by the primary purpose for which they were collected, e.g. revenue from rates and taxes) and often out of date (because of the low priority given to their compilation). Improving official statistics depends on the importance given by government departments to this function and on the education of the public. These factors change slowly in Africa, and indeed in most developing countries, and often independent surveys are required to confirm or show the shortcomings in official statistics to plan their improvement and to supplement them. Most African countries therefore depend on censuses and sample surveys designed specifically to meet their data needs.

However, socioeconomic data collection, analysis and research also suffer in Africa more often than not from several severe limitations. First, there is the problem of ethnocentricity: that is, an obsession with modernisation and a restriction of the scope of the investigation to the point where the questions supply their own answers. In addition, analysis from the economic standpoint is rendered difficult by data that are often much poorer than the analyst realises. Caldwell (1977) gave interesting examples of several of this type of limitation in studies on populations of African countries; for example, he noted that:

> The type of fertility research that attempts to provide social and economic explanations for fertility levels and trends has arisen almost entirely from Western quantitative sociology and very largely from American sociology. Many of these practitioners in the Third World have had relatively short periods of residence in the countries in which they undertake research, and usually none in the rural areas, and they work with nationals of these countries who were as often as not their students or colleagues in the universities from which they came. Inevitably, concepts are taken from the West to the culture being studied. Often concepts are sufficiently inapplicable seriously to distort the findings.
>
> The most harmful distortion probably relates to the nature of the

> family both as a social and economic unit. Western sociologists are capable of perceiving that the nuclear (or conjugal) family of the West, with an economic concentration within a group made up of husband, wife and non-adult children and usually residence restricted to the same group, is not universal – that traditionally it may have been a rarity elsewhere. But when carrying out social surveys outside the West they tend to treat other types of kinship as if the differences were restricted to having financial responsibilities towards a few more relatives – often assuming that these responsibilities must be more reluctantly borne than those towards members of the nuclear family – or living with a few more of them. They find it difficult to appreciate that a married man might prefer to spend his earnings on his brother rather than his wife, that he might have budgetary arrangements more in common with his father than with his wife, that he might feel that his uncle comes within the circle of his immediate family of affection and preferred expenditure as much as his children. In sub-Sahara Africa, parts of Asia and undoubtedly elsewhere, the nuclear family is usually the creation of the research worker, and further economic analysis based on such a unit is quite invalid.

Second, governments in African countries have at various times interfered with data collection and analysis, thereby contributing to the various problems connected with the quality of the data and their reliability. This can arise in several ways (see, for example, Dasgupta and Seers, 1975). First, governments can ask the statistical office to manipulate or fabricate data. In the past, census operations of several countries have been suspected of being manipulated to exaggerate the proportional share of particular ethnic, racial or religious groups or regions. Nigeria provides an excellent example in this category, where both the 1963 and the 1973 census ended in dispute. Udo (1978) gave four main reasons why the 1973 census failed, one of which was the inflation of census figures in the field. (For further discussion of these points, see, for example, Udo, 1968 and 1978.) To choose 1963 census figures as Nigeria did in favour of the 1973 census data, amounted to a choice between two unreliable census figures for the purposes of planning. A more regular practice, however, is to produce a rosy picture of the future, with long-term statistical forecasting based on dubious assumptions and data in order to induce the people to vote for, or otherwise support, the government.

Second, the statistical office can be requested to suppress inconvenient data. This is a regular practice in many countries. The politicians in power justify this on the grounds of public policy.

Third, in several countries, while the government would not favour outright suppression of data, attempts are made to lessen the impact of unfavourable data by inducing the statistical office to change the format of statistical tables to hide some individual items under aggregate figures, or to hold back publication for some time until after an election or an improvement in the economic position, or to add explanatory notes to tables to convey a false interpretation of the figures.

Political bias can also enter into statistical work in many other forms: for example, through the appointment of the head of the statistical office; by encouraging research in certain directions and by withdrawing support for research on potentially unfavourable topics, such as poverty or income distribution; and by using data on individuals and firms for the purpose of political blackmail.

Finally, sometimes as a result of pressure from international and, in some cases, bilateral agencies on national statistical offices to fill in returns or to provide information to back an aid request, estimates have been produced that are neither needed by local policy-makers or analysts nor of defensible quality professionally.

As a result of these activities, one may ask: how reliable, then, are the published statistics in Africa? Reliability has a number of different dimensions. There is a whole range of definitions, and behind the different interpretations of the figures there is a range of values – emotional, political and ideological. There is hardly any politician who is consistent from year to year in his definition of unemployment, if he ever tries to define it at all. But the crucial questions are: Are most statistical figures so vague that they can be turned and twisted to support ideologies and views? Have statistics become a tool in the hands of those whose political or economic interests diverge from a meaningful development of Africa? (For an interesting discussion of issues raised by this question, see Rondos, 1979.) Probably not, but too often the wrong figures are chosen at the wrong time, at the wrong place, for the wrong reason, for the wrong purpose, by the wrong person (Bondestam, 1973). There is usually a range of data to choose among and to combine in various ways, and it is always possible, if backed by a sufficient degree of dishonesty, to find some combination of figures that suits even the most odd purposes and supports even the most odd ideas.

Statistically, the most stringent test of reliability is often the extent to which the results of an inquiry as first published are revised to take account of errors previously undetected in the figures received earlier, or of errors which were made in compiling the statistics, or of changes between figures imputed for a return not received by the closedown date and the actual figures that the respondent supplies later, or – and this is the most common source of revision in the results of a new inquiry – because data were incomplete and investigations lead to additions (Fessey, 1978).

In general, of course, the degree of reliability decreases with the distance to the source; for example, a serious research report gives more reliable data than second-hand quotations from that report, because there is a tendency to smooth out figures or round them off, to exaggerate their validity, to forget definitions, to ignore or misunderstand important facts, or simply to quote out of context when given second hand. An estimated population growth rate of 2·7 per cent per annum from a serious report may end up being quoted as 'about 3·0 per cent per annum' or as 'less than 3·0 per cent per annum'. Examples are not difficult to find, and this is given to support the need, if not the necessity, to go back to the original source and to check the definitions used behind the statistical data.

What, then, is to be done to remedy a situation in which the supply of statistics still seems to have so many shortcomings. As users, we cannot do very much more than what is being done about inaccuracy by producers. To use statistics means accepting, unconsciously for the great majority of users, the judgements that lie behind them, especially those which have influenced the documentation of one sort of activity and the exclusion of another. All that one can ask for is that the professional players in the game stick flags on such data saying 'user beware', while not estimating measures that are crude to more than two significant figures. Suitability of data for use, therefore, will have to be judged against the various stages in the collection procedure and its reliability assessed.

Another part of the answer, as ever, must lie in better training in the use of statistics, but there are some major difficulties. In the first place, when it comes to commercial or economic decision-making, statistics are only part of the input. They are necessarily mixed with guesswork, direct observation, lies and inside information. The only adequate training is the hard school of experience, which creates the foundation of intuition upon which, in the end, we must rely.

In addition to what we have said, it is important to observe some basic guiding rules in the utilisation of the data. First, always give the source of the data. If the reader is suspicious about the reliability of the statistics, he can always cross-check this with the original source. Unless the source is mentioned, the suspicion may be strengthened, and the validity of the conclusions may be questioned.

Second, attempts must always be made to find the possible sources of error and some idea about the reliability of the data.

Third, do not be satisfied with averages for the whole population; these can be misleading. Always remember that aggregate statistics are only as good as the components of which they are constituted; therefore, check the national averages with regional and other variations in the data.

Fourth, do not always accept exact figures as being better than approximate ones. For example, to say that the population of Sierra Leone in 1963 was 2,180,101 does not give more information than to put the figure at 2·2 million; in other words, figures that are essentially broad brush should not be treated as fine grain. In this example, the figure is probably biased and may have been underenumerated or overenumerated anyway.

Fifth, if growth figures are given over longer periods, recalculate them to be valid on a yearly basis. To say that a country's industrial growth was 61 per cent during a five-year period may lure the uncritical reader to believe that the yearly growth rate was slightly above 12 per cent (61/5 = 12·2), whereas it actually was 10 per cent ($1.1^5 = 1{\cdot}61$).

Finally, recorded statistical data can, in certain cases, be improved by various adjustment procedures, say age data from population censuses or surveys. However, such advanced calculations usually demand a fairly good knowledge in the respective field (economy, demography, and so on) and statistics in general. Statistical data with doubtful reliability should not be stretched or extended to give new figures, which can never be better than the original ones or be used for other lengthy and complicated calculations. Even small errors in quantitative data may lead to completely wrong judgements.

CONCLUDING REMARKS

The value of statistics is judged by their usefulness in social and economic development and policy formulation. In Africa, however, their value is questionable, because they are subject to omissions and misreporting. For comparative analysis, additional problems are introduced by the use of non-comparable definitions in data collection procedures and by differences in the detail of tabulation. Given the incomplete, fragmentary and defective nature of the data available, their interpretation must remain somewhat tentative, and an uncritical acceptance of a single set of data may be very misleading. Nevertheless, the importance of data in formulating national development policies and objectives requires a reappraisal of the data needed to help to formulate, monitor and evaluate these policies. This demands a critical examination of the relevance, adequacy and reliability of current statistical data and of the tools used to collect and analyse these data.

Although there are existing collections of statistics on Africa, these are very uncritical about the techniques and conceptual problems with the figures that they present. Knowledge of the limitations and their reliability is an indispensable condition for a meaningful utilisation of African statistics. Accordingly, the main theme of this book is to discuss some of the most common weaknesses, problems and difficulties of African statistics. It will focus on three main aspects: problems connected with the collection of statistics and their consequences for the reliability of statistics; the reliability of some of those statistical measures which are most frequently used in the description of the social and economic conditions in Africa, and some advice for the interpretation and utilisation of these statistics; and finally, how the weaknesses that characterise some of the statistical measures can be avoided. The broad purpose is to provide guidance to what is currently available, both to facilitate access and to forewarn the unwary about deficiencies and pitfalls in source material on social and economic statistics.

REFERENCES AND FURTHER READING

Allen, Sir R. (1970), 'On official statistics and official statisticians', *Journal of the Royal Statistical Society*, series A, vol. 133, pp. 509 – 29.

Bondestam, L. (1973), *Some Notes on African Statistics: Collection, Reliability and Interpretation*, Research Report No.18 (Uppsala: Scandinavian Institute of African Studies).

Caldwell, J. C. (1977), 'The economic rationality of high fertility: an investigation illustrated with Nigerian survey data', *Population Studies*, vol. 31, no. 1, pp. 5 – 28.

Dasgupta, B. and Seers, D. (1975), *Statistical Policy in Less Developed Countries*, IDS No.114 (Brighton: Institute of Development Studies, University of Sussex).

Economic Commission for Africa (ECA) (annual), *African Statistical Yearbook* (Addis Ababa).

ECA (annual), *Demographic Handbook for Africa* (Addis Ababa).

ECA (irregular), *Statistical and Economic Information Bulletin for Africa* (Addis Ababa).

Fessey, M. C. (1978), 'Business statistics: a decade of central collection (with discussion)', *Journal of the Royal Statistical Society*, series A, vol. 141, pt 4, pp. 478 – 506.

Miller, J. A. (1977), 'Statistics: a city view', *The Statistician*, vol. 26, no. 4, pp. 233 – 42.

Omaboe, E. E. (1963), 'Some observations on the statistical requirements of development planning in the less-developed countries', *Economic Bulletin of Ghana*, no. 2.

Rondos, A. (1979), 'Ivory Coast: whose success?', *West Africa*, no. 3222 (16 April), pp. 657 – 9.

Stafford, J. (1968), 'The development of industrial statistics', *Statistical News*, 1, pp. 1 – 10.

Udo, R. K. (1968), 'Population and politics in Nigeria', in J. C. Caldwell and C. Okonjo (eds), *The Population of Tropical Africa* (London: Longman).

Udo, R. K. (1978), 'Thinking of a census before 1983', *Daily Times of Nigeria* (13 May).

United Nations (1971), 'Demographic data collection', paper prepared by the UN Statistical Office for the African Population Conference, Accra (9 – 18 December), No. E/CN.14/POP/51, annex 1.

United Nations (annual), *Statistical Yearbook* (New York: UN Statistical Office).

United Nations (annual), *Demographic Yearbook* (New York: UN Statistical Office).

Wilson, H. (1973), 'Statistics and decision-making in government: Broadshaw revisited', *Journal of the Royal Statistical Society*, series A, vol. 136, pp. 1 – 20.

MAJOR PUBLICATIONS ON SOCIAL AND ECONOMIC STATISTICS

ALGERIA

Service de la Statistique Générale (Algiers):
- *Bulletin de statistiques générales.*
- *Note de conjoncture.*
- *Annuaire statistique de l'Algérie* (annual).
- *L'Algérie en quelques chiffres* (annual).

Ministère de Commerce:
- *Rapport de synthèse de l'activité économique d'après les rapports des Chambres de Commerce et d'Industrie* (monthly).

ANGOLA

Repartição de Estatístics Geral.
- *Anuário estatístico de Angola* (annual).
- *Boletin mensal de estatística* (monthly).

BENIN (formerly Dahomey)

Direction de la Statistique (Coutonou):
- *Bulletin de statistiques.*
- *Bulletin économique et statistique.*
- *Annuaire statistique* (annual).

Direction générale des affaires économiques, Aspects économiques.

Ministère du Commerce de l'économie et du tourisme, Bulletin économique et statistique.

BOTSWANA

Central Statistics Office (Gaborone):
- *Statistical Abstract.*
- *Statistical Newsletter.*

BURUNDI

Direction de la Statistique (Bujumbura):
- *Bulletin de la statistique.*
- *Annuaire statistique: supplement au Bulletin de la statistique.*
- *Bulletin trimestriel.*

CAMEROUN

Direction de la Statistique (Yaounde):
- *Annual Report of West Cameroun.*
- *Monthly Digest of Statistics.*
- *Statistical Annual Report of West Cameroun.*
- *Resumé des statistiques de Cameroun oriental.*
- *Bulletin mensuel de statistiques.*
- *Note trimestrielle sur la situation économique.*
- *Quarterly Bulletin of Statistics* (Department of Statistics and National Accounts)
- *Note trimestrielle de statistiques* (series A and B).

Regional Statistical Service, *Statistics Annual Report of West Cameroun* (Buea).

CENTRAL AFRICAN REPUBLIC

Direction de la Statistique (Bangui):
- *Bulletin mensuel de statistiques (nouvelle serie).*
- *Bulletin mensuel de statistiques.*

CHAD

Service de la Statistique (N'Djamena):
- *Annuaire statistique de Tchad* (annual).
- *Statistiques et commentaires* (annual).
- *Bulletin mensuel de statistique.*

EGYPT
Central Statistics Agency (Cairo):
Statistical Abstract of the UAR.
Statistical Handbook.
Annuaire statistique.

ETHIOPIA
Central Statistical Office (Addis Ababa):
Statistical Abstract (annual).
Statistical Pocket Book (annual).
Statistical Bulletin (last one no. 9: *Urbanisation in Ethiopia*).

GABON
Service National de la Statistique (Libreville):
Bulletin mensuel de statistiques.
Etudes et enquêtes statistiques: supplement au Bulletin mensuel de statistiques.
Situation économique, financière et sociale (annual).
Annuaire statistique.

GAMBIA
Central Statistical Office, *Statistical Summary* (Banjul).

GHANA
Central Bureau of Statistics (Accra):
Quarterly Digest of Statistics.
Monthly Economic Bulletin.
Bureau News (monthly).
Economic Survey (annual).
Statistical Yearbook (annual).
Statistical Handbook of the Republic of Ghana.

IVORY COAST
Direction de la Statistique (Abidjan):
Supplement trimestriel au Bulletin mensuel de statistiques.
Bulletin mensuel de statistiques.
Situation économique de la Côte d'Ivoire.

KENYA
Economics and Statistics Division (Nairobi):
Statistical Digest.
Economic and Statistical Review.
Economic Survey (annual).
Statistical Abstract.

LESOTHO
Bureau of Statistics *Annual Statistical Bulletin* (Maseru).

LIBERIA
Department of Planning and Economic Affairs (Monrovia):
Annual Report.
Economic Survey (annual).
Bureau of Statistics, *Quarterly Statistical Bulletin of Liberia.*

LIBYA
Census and Statistical Department (Tripoli):
Statistical Abstract.
Economic Bulletin.
Quarterly Bulletin of Statistics.

MADAGASCAR
Institut National de la Statistique et de la Recherche Economique (INSRE) Tanararive):
Bulletin mensuel de statistique.
Situation économique.

MALAWI
National Statistical Office (Blantyre):
Annual Survey of Economic Activities.
Monthly Bulletin of Key Economic Indicators.
Economic Planning Division, *Economic Report* (annual).

MALI
Service de la Statistique générale (Bamako):
Bulletin mensuel de statistique.
Annuaire statistique (annual).

MAURITANIA
Direction de la Statistique (Nouakchott):
Bulletin mensuel statistique.
Abrégé de statistique (bi-annual).
Annuaire statistique (annual).

MAURITIUS
Central Statistical Office: (Rose Hill)
Bi-Annual Digest of Statistics.
Quarterly Digest of Statistics.

MOROCCO
Direction de la Statistique (Rabat), *Bulletin mensuel annuaire statistique du Maroc.*

NIGER
Ministère des Affaires économique et de l'Industrie, *Bulletin mensuel d'information économique* (Naimey).
Service de la Statistique, *Bulletin trimestriel de statistiques.*

NIGERIA
Federal Office of Statistics (Lagos):
Digest of Statistics (quarterly).
Annual Abstract of Statistics.
The Economic Indicators (monthly).
Digest of Local Government Statistics: Ibadan.
Western Statistical Abstract (half-yearly).
Western Region: Supplementary Estimates (annual).

PEOPLE'S REPUBLIC OF CONGO
Service National de la Statistique (Brazzaville):
Bulletin mensuel de statistiques.
Annuaire statistique.

SENEGAL
Direction de la Statistique (Dakar):
Bulletin statistique et économique (monthly).
Situation économique du Sénégal.

SIERRA LEONE
Central Statistics Office (Freetown):
Quarterly Statistical Bulletin.
Economic Review.
Annual Statistical Digest.

SOMALIA
Statistical Department (Mogadishu):
Somalia Statistics (monthly).
Monthly Statistical Bulletin of the Somalia Republic.
Statistical Abstract of Somalia.

SUDAN
Department of Statistics (Khartoum):
Economic and Financial Bulletin (quarterly).
Economic Survey.

SWAZILAND
Department of Statistics (Mbabane):
Statistical News.
Annual Statistical Abstract.
Annual Statistical Bulletin.
Statistical News and Economic Indicators.

TANZANIA
Central Statistical Bureau (Dar es Salaam):
The Economic Survey.
The Economic Survey and Annual Plan.
Monthly Statistical Bulletin.
Quarterly Statistical Bulletin.

TOGO
Direction de la Statistique (Lomé):
Bulletin de statistiques (monthly).
Annuaire statistique de Togo.

TUNISIA
Institut de Statistiques (Tunis):
Bulletin mensuel de statistique.
Annuaire statistique de la Tunisie.
Revue tunisienne d'économie et de statistiques (annual).

UGANDA
Statistics Division (Entebbe):
Statistical Abstract.
Quarterly Economic and Statistical Bulletin.

UPPER VOLTA
Direction de la Statistique, *Bulletin mensuel d'information statistique et économique* (Ouagadougou).

ZAIRE
Institut National de Statistique, *Bulletin trimestriel de statistiques générales* (Kinshasa).
Office National de la Recherche et du développement, *Notes de statistiques et de conjoncture.*

ZAMBIA

Central Statistical Office (Lusaka):

Monthly Digest of Statistics.

Annual Report: Department of Census and Statistics.

Statistical Yearbook (annual).

East African Statistical Department, (Nairobi).

Quarterly Economic and Statistics Review

CHAPTER 2

Population Statistics

Reference has been made in Chapter 1 to the nature of statistics, methods of their collection and their deficiencies. One important area where data collection is essential is in the field of population, whose study generally requires extensive and accurate statistics. The collection of the relevant material is a recent development in Africa. Although the data are generally defective and the costs of collection, analysis and publication of such data expensive,[1] costs can be justified in economic terms in a country that can use the results valuably for administrative, social and economic planning purposes. This chapter gives the sources of demographic data, their uses and limitations.

SOURCES OF POPULATION DATA

National Sources

(1) *Population Censuses*
The census is the main source of demographic statistics in many countries. It is usually defined as: 'The total process of collecting, compiling and publishing demographic, economic, and social data pertaining at a specified time or times to all persons in a country or delineated territory' (United Nations, 1965). Unfortunately, population censuses are usually undertaken after a long span of time, not to mention the fact that the processing also takes a long time because of the great volume of data involved. When these much needed statistics on population are finally available, they may no longer be useful in the assessment of current population changes. Changes in population characteristics would necessitate readjustment of national development plans. The statistics are published in special census reports or volumes of various countries. The frequency of publication naturally varies from country to country.

(2) *Vital Registration*
This is another source of population statistics. It consists of continuous registration of births, deaths, marriages, and so on. In many developing countries, vital statistics are non-existent; in those where they exist, the data are inadequate and defective to the extent

that they can hardly serve to measure accurately levels of fertility and mortality and their trends at any given time. The building up of a reliable vital registration is both an expensive and a protracted process. As a general rule, the data are published in vital registration reports monthly or annually in various countries.

(3) *Sample Surveys*
A sample survey is the third source of demographic statistics, which seeks to collect information from a fraction of the population only. It is employed to arrive at estimates of demographic characteristics, size, distribution, mortality, fertility and migration. The data from this source alone are not sufficient to meet the heavy demands for various population statistics because of financial and technical constraints. The statistics are issued in special reports.

(4) *Population Registers*
This source is usually employed by the developed countries, which maintain a regularly updated list of people resident in a country with details of sex, date of birth, marital status, and so on. The maintenance of population registers demands a reasonably accurate address system and a literate population. The financial costs involved in maintaining the system are formidable.

(5) *Non-Traditional Sources*
These include parish registers, baptismal records and other administrative data.

International Sources

The United Nations *Demographic Yearbook*, first published in 1948 by the United Nations Statistical Office in New York, furnishes internationally comparable and comprehensive annual statistics on population – its size, fertility, mortality, urbanisation, and so on – which are put to a wide range of uses by governments, institutions, private organisations and international agencies. The *Yearbook* occasionally puts out special editions covering in detail specific topics: for example, mortality and fertility. The data in the *Yearbook* are taken either from information supplied to the office of the United Nations by national statistical departments or offices or from official publications. Annotated citations to the official demographic statistics of all countries, as well as other books and articles on population, appear in the *Population Index*, a quarterly publication by the Princeton University Office of Population Research.

USES OF POPULATION DATA

Uses of Census Statistics

Population censuses are the primary sources of the basic data required for planning, for administration and also for many aspects of economic and social research. Their uses are enormous.

(1) *Educational Planning*
Population plays a role on the supply side in educational planning; the population in the school-going ages by age and sex must be available. To plan for the number of school places, the number of teachers and the characteristics of the educational system – that is, student/teacher ratios and student/classroom ratios (called technical coefficients) – are required. The needs for teachers and classrooms at various levels of education, given the enrolment targets, can be projected, as well as the population in the school-going ages at future dates.

(2) *Planning for Housing*
The demand for housing is clearly tied to the population size and its age–sex distribution, the rate of household formation and dissolution, migration and population redistribution.

(3) *Agricultural Planning*
Relationships between demographic changes and agricultural development must be among the most important and wide-reaching interactions between population and socioeconomic change. Most of the population in developing countries resides, works and consumes in agricultural areas, and the agricultural sector plays an important role in overall economic development as a source of food for the entire population. Governments therefore want to know how many people work in agriculture, their age–sex distribution, their dependents, the income per head of agricultural workers, and so on. All these data can be obtained from agricultural censuses or population censuses.

(4) *Regional Planning*
Rural–urban migration accounts for almost half of the urban population growth in most countries, and most governments have expressed concern at the effect of migration, particularly of urban population growth. Migration and regional social and economic planning are intimately related, since migration alters the spatial distribution of the population and thereby influences the location of

government services. Data from censuses on both the levels of migration and the characteristics of the migrant are needed for appraising the consequences of migration on both sending and receiving areas.

The study of migration is also important for other reasons apart from the social and economic consequences. In demographic analysis, for example, migration is considered as one of the most complex of the basic demographic variables. Like death, migration can take place at any age. Migration and natural increase together form the two components of population growth and redistribution, but their effects are not always synchronous, nor are they mutually exclusive. Migration differs from mortality and fertility because each migration event has a separative and additive aspect simultaneously. In other words, the contribution of migration may be either positive or negative, but the contribution of natural change to area growth is usually positive.

(5) *Manpower and Labour-Planning*

Employment is one of the most important social and economic problems in developing countries; and for the government to plan for the provision of adequate employment for persons in the labour force, it needs to know how many persons are in the labour force, the rate at which new persons are entering and leaving the labour force and the characteristics of the labour force. The goal of manpower-planning is to project labour force supply and demand and to make recommendations for government policy to help reconcile the two. The size of the future labour force depends upon the projected population and labour-force participation rates. The projected population, together with its age – sex distribution, determines the population of labour force age. Since both the projected population size and its age distribution depend primarily upon future fertility levels and only secondarily upon mortality levels, fertility rates are crucial determinants of long-run labour supplies.

(6) *Health-Planning*

The increasing desire of governments to improve health conditions and to engage in health-planning – that is, to plan for the number of persons per doctor and nurse, the number of persons per hospital bed, and so on – requires demographic data of the total number of persons in the country and the age – sex distribution, since different sections of the age span make different demands on the health services.

Summarising, demographic data are required: (1) for estimating the future population sizes and distribution, which is fundamental to long term planning of many public programmes, educational needs (the number of school places and teachers), health needs (the number and location of hospitals), manpower needs (the employment of the population), and so on; (2) for investigating adequate relationship between demographic and socioeconomic processes; (3) for measuring the levels and trends in the standard of living of the population; (4) for the allocation of parliamentary seats; and finally, (5) for the allocation of resources.

Uses of Sample Surveys

In Africa, the history of sample surveys as a source of demographic statistics, unlike censuses, is relatively recent, spanning only about twenty to twenty-five years. These surveys were initiated first in the French-speaking African countries and later in other former British colonies. The main objectives are: (1) the estimation of the total population and its geographic distribution; (2) the checking of the accuracy of a population census; (3) the estimation of the components of population change (fertility, mortality and migration); and (4) the estimation of a number of demographic characteristics of the existing population.

Of these four objectives, the second needs further elaboration. This is the technique of postenumeration surveys. The method has become one of the most important survey techniques for checking the accuracy of census enumerations. The fact that a small well-designed and well-administered sample survey can produce more accurate measurements than is feasible in the much larger scale operations required for a census, suggests the use of the former as a vehicle for checking on the latter. This survey is always conducted after the main census. It is also usual to conduct the survey as closely as possible to the date of the census so that the two sets of data collected are comparable.

As the results from a postenumeration survey are meant to serve as a standard against which to judge the figures obtained in the census, it is very important to take all possible steps to ensure that the survey is reasonably free from errors. This is usually done by employing in the survey interviewers whose performance in the census has been judged as excellent.

One last point concerns sampling as an aid to censuses. Censuses are an invaluable aid to sample surveys because they provide a frame from which the sample can be selected and furnish benchmark data which can be used for increasing the precision of the sample estimates. However, both censuses and sampling methods are interdependent.

Sampling methods too can play an important part in the conduct of the census. In the first place, the scope of the census can be broadened by asking certain questions of only a sample of the population. This reduces the burden of response and the time required for carrying out the work. Second, the important results of the census can be tabulated speedily on the basis of a sample of returns. In the third place, the census can be taken on only a sample of areas or other units in order to save costs, to save time or to provide information under conditions that render a complete enumeration impossible. Finally, the errors in processing the census returns can be kept within reasonable bounds by examining a sample of returns at various stages of the processing.

Uses of Vital Registration Records and Statistics

The need for establishing a vital registration system, or for improving the quality and quantity of vital statistics in existing schemes, depends to a considerable extent on the administrative and statistical requirements in various countries. It demands first of all a demonstrated use for the product of the functions of such a system; this can only come about when the uses of birth, death and marriage certificates, and so on have become essential in the social and economic organisation of the communities, for education, for participation in adult life by voting, for state support in old age and for inheritance. These uses for both records and statistics must be important enough to justify government action to meet these needs and to institutionalise the system.

Specifically, there are three main categories for the use of vital records and statistics:

(1) *Administrative Uses*

(a) Birth registration records provide the necessary legal and documentary evidence to certify a person's age, parentage, birthplace and nationality or citizenship.

(b) Birth registration certificates are used in the determination of a person's eligibility for admission to school, to obtain a passport, to enter certain fields of employment and to vote in an election.

(c) In the exercise of other civic functions, such as entitlement to family allowances, insurance claims, care of children, tax deductions or benefits, the birth certificates will be required.

(d) The records of death (death certificate) provide legal evidence relevant to claims of property, to insurance on deceased persons, to the rights of surviving spouses to remarry and to claims for family allowance where the death creates financial need.

(e) Marriage and divorce records are the basis for claims involving the status of women – such as dependency, alimony payments, tax deductions and allowances – and the provision and allocation of specific types of housing. The divorce records are also important to establish the right to remarry.

(f) In the initiation and maintenance of population registers, officially authenticated records of birth, death, marriage and divorce have become essential and invaluable because changes in the physical, civil and geographical status of the family enter the population register through records of birth.

(2) *Statistical Uses*

(a) Vital statistics provide additional data, independent of census, on measures of fertility and mortality both for the entire nation and for small geographical areas for study of trends and patterns in fertility and mortality.

(b) The sex ratio at birth provides a very important piece of information for use in population projections.

(c) The data from vital registration gives information on the seasonality of birth and death; this has relevance to the planning of related services.

(d) The system is used as a check on census enumeration, particularly at the infant and young childhood age range, where under-enumeration is common.

(e) In populous countries, vital statistics from a vital registration system can be useful in formulating population control programmes and for evaluating the effectiveness of such programmes.

(f) In a longitudinal-type survey on genetic studies about fertility and mortality, records of birth, death, marriage and divorce can be very useful.

(3) *Medical Uses*

(a) Data about deaths and their causes as recorded on death certificates are essential in the planning of health services for specific areas as well as for the whole country; for example, the cause-of-death statistics are helpful in forecasting the future course of age-specific mortality.

(b) Epidemiological studies based on causes of morbidity, which have a strong influence in the reduction of mortality, are usually based on records of vital statistics generated by the system.

Given a dynamic situation in both the social and economic development in the world today, many uses of vital records and statistics can be expected to emerge in the not-too-distant future.

Uses of Population Registers

The population register is basically a regularly updated list of people resident in a country, with details of sex, date of birth, marital status at least, and information on current area of residence. Population registers vary greatly in many respects, but their uses are similar. These uses may be grouped into two classes: administrative and statistical.

(1) *Administrative Uses*
Administrative uses of registers include identification of persons, control of electoral rolls, control of selection for military service, indications of each person's status with respect to various social security benefits and the preparation of tax lists. Registers can also be used for the control of food-rationing in times of shortage and for the implementation of regulations concerning compulsory education.

(2) *Statistical Uses*
The register can be used as a source of important statistical information. The most important of the many statistical applications of population registers are: population statistics; vital statistics; size and characteristics of the population; population changes; migration statistics (internal and external); electoral statistics; social security statistics; health statistics; judicial statistics; educational statistics; statistics of family income and consumption; labour statistics; income tax statistics; family and housing statistics; statistics for follow-up studies in demographic, educational and medical fields; comparison of population register data with census data; preparation of population or subpopulation lists by address, age groups, sex, and so on; and finally, the use of the register as sampling frames.

The main advantages of the registration method are temporal in nature; only with continuous registration is a complete historical time series possible. Time series, so derived, have a number of analytical uses. They can, for example, be correlated with economic time series. However, special care must be exercised in such correlations to free the register series from accumulations of errors and to distribute any discrepancies that may be revealed by periodic external checks. The fact that errors in data collected in this manner are cumulative can result in greater bias in population register data than one is likely to encounter in data from other sources.

A principal difficulty is the maintenance of accurate address details. Even where registration of change of address is a legal requirement, compliance may be difficult to ensure. For statistical purposes, some degree of error may be tolerable, but it needs to be carefully assessed. The population register can provide a framework for the development of record linkage systems for statistical purposes, using a system or systems of personal reference numbers to facilitate linkage. The practical difficulties in introducing population registers in Africa should, however, not be underestimated, but their potential for demographic analysis is so great that an experiment in the maintenance of the register is worth trying in a number of countries. The nearest that African countries have got to population registers is the implementation of the administrative census, which is the national total of populations registered by local administrative officers in the Francophone countries of the continent.

Uses of Non-Traditional Sources

Administrative censuses were introduced by the French and Belgian colonial systems, whereby every year, or at a given period, administrative officers went to villages and recorded in a register the number of persons living there in order to tax the villagers and to draft young men into the army. As a result of this, the total number of persons registered, counted or recorded is taken as the barest minimum, and serious underenumeration can be suspected. Consequently, the estimates of annual growth rate can also be regarded as underestimated for reasons inherent in the very motives underlying the enumeration. Nevertheless, some attempts have been made to exploit the available data in order to estimate annual growth rates and examine possible trends. Specific uses of this type of data are:

(1) The processing of administrative census registers helps to provide information on the size of the population and its distribution by age and sex and by social, professional and occupational groups.
(2) It enables the verification or matching of data obtained from a survey with records of school registers or baptismal records.
(3) It enables the provision of a historical view of demographic phenomena over time.

There are three advantages in using administrative data for statistical purposes: (1) they may be more accurate than statistical returns; (2) they are on a 100 per cent rather than a sample basis; and (3) they do not impose an additional burden on the public. There are disadvantages too: (1) the concept and definitions devised for administrative schemes may not suit statistical needs; (2) the information may not cover the whole of the relevant population; and

(3) many of the records may be out of date, particularly where individuals have only infrequent and irregular contact with the administrative system.

AVAILABILITY OF POPULATION DATA IN AFRICA

The scope, nature and availability of censuses and other demographic-data collection systems throughout the world vary considerably. The major over-riding criteria for deciding on the number and type of questions included in censuses have been, and still are, the needs of the various countries in the field of demographic data collection for planning purposes, and the availability of resources (both financial and human) to carry through the operation. The latter factor has affected developing countries considerably and has in fact limited the scope of most censuses in Africa. A secondary requirement is the need to achieve worldwide comparability in concepts and topics for inclusion in census programmes. These needs have, however, changed from time to time, but it is possible to come to some agreement on the general and minimum needs for the various parts of the world.

The availability of demographic information from censuses and surveys is very uneven in its coverage. Certain areas and population groups, notably nomadic peoples, are excluded altogether from several demographic-data series. Differences between countries also occur in the definition of the population to which the collected data refer; for Francophone areas this is generally the *de jure* population; for other areas it is often the *de facto* population. Comparability of data is limited by (1) the staggered timing of data collection, and (2) the detail of the published tabulations. On the whole, African countries are still not collecting meaningful data on employment, education and migration, among other things. Part of the reason is that the census officials have not been exposed to as much demand and advice in these areas as they have in the areas of fertility and mortality. Table 2.1 gives some basic information on dates and types of censuses and surveys taken in recent years in Africa.

LIMITATIONS AND QUALITY OF POPULATION DATA

Each of the traditional methods of demographic data collection suffers from some type of shortcoming. These deficiencies can be observed, particularly when existing methods are measured against the types of population data requirements relating to (1) data content, (2) geographical coverage, (3) population composition, and (4) time factors.

Table 2.1 *The major censuses and surveys in subSaharan Africa in the 1950s, 1960s and 1970s*

Country	Type of Inquiry			Area Covered				Status of Work	
		Sample inquiry							
	Census	Single round	Multi round	Whole country	Urban	Rural	Year of Inquiry	Published	Other[a]
Angola	x						1950	x	
	x						1960	x	
	x						1970		x
Botswana	x						1954	x	
	x						1964	x	
	x						1970		x
Burundi		x		x			1952-7	x	
		x		x			1965	x	
			x	x			1970-1		x
Cameroun (East)		x		x1			1960-1	x	
		x		x2			1962-3	x	
		x		x3			1962	x	
		x		x4			1965	x	
Cameroun (West)		x			x		1964	x	
Central African Republic		x		x			1959-60	x	
		x		x			1960-1	x	
	x					x	1961-3	x	
Congo Democratic Republic		x		x5			1958-9	x	
		x		x6			1960-1	x	
Dahomey (now Benin)		x		x			1961	x	
Ethiopia		x		x			1964-7		
		x		x			1970		x
Gabon	x	x		x			1960-1	x	
			x				1962		x7
			x	x8			1970		x
Gambia	x			x			1963		x
Ghana	x			x			1960	x	
		x9		x			1960	x	
			x10	x			1968-9	x	
	x			x			1970	x11	
		x	x				1971		

Country	*Type of Inquiry*			*Area Covered*				*Status of Work*	
		Sample inquiry							
Country	*Census*	*Single round*	*Multi round*	*Whole country*	*Urban*	*Rural*	*Year of Inquiry*	*Published*	*Other*[a]
Guinea Bissau	x			x			1950	x	
	x			x			1960	x	
Guinea Republic		x				x12	1954-5	x	
Ivory Coast		x		x			1957-8	x	
			x		x		1963-4	x	
		x				x	1962-5	x	
Kenya	x	x		x			1962	x	
	x			x			1969	x	
Lesotho	x			x			1956	x	
	x			x			1966	x	
Liberia	x			x			1962	x	
			x13				1969-73	x	
Madagascar (now Malagasy Republic)			x			x	1967-8	x	
			x			x	1969-70	x	
Malawi	x			x			1966	x	
			x	x			1969-70		x
Mali		x		x			1957-8	x	
		x		x			1960-1	x	
Mauritania		x			x		1961-2	x	
		x		x			1964-5	x	
Mozambique	x			x			1950	x	
	x			x			1960	x	
	x			x			1970		x
Namibia	x			x			1970		x
Niger		x14		x			1959-60	x	
		x15		x			1963	x	
Nigeria	x			x			1952-3	x	
	x			x			1962-3	x	
			x			x	1965-6	x	

Notes:

a - Status unknown

x1 - Northern Region

x2 - South-Eastern Region only

x3 - Central and Eastern Regions

x4 - Bamileke Region
x5 - Kouilou Region
x6 - Ten rural districts
x7 - Publication abandoned
x8 - One region
x9 - The 1960 post enumeration survey
x10 - The dual system method
x11 - Volume 2 published
x12 - Four rural districts
x13 - The dual system method used in the population growth estimation
x14 - Sendentary population
x15 - Nomada

Source: Wogugu (1974).

Errors in Census Data

The accuracy of census information has varied from time to time and from place to place all over the world, but some generalisations are possible about the types of errors found in censuses in Africa. Two main types of error are common: errors of coverage (under-enumeration or overenumeration) and errors of content. In coverage errors, villages, hamlets, households, persons, and so on may be missed or not counted at all; this is known as undercount or underenumeration. If households, persons or villages are enumerated or counted more than once, a duplication or an overenumeration occurs. On the other hand, if a unit is enumerated but there is an error in the recording of its relevant characteristics — for example, in the recording of the age of a person — a classification or content error is made. This type of error is more common in censuses of developing countries than in those of developed countries.

Age errors are made on a vast scale and are usually partly systematic. The main types of age errors are: (1) age-heaping (digital preference); and (2) age-shifting, which often occurs in certain sections of the age span more than in others, say the older age groups. The former is a process whereby the respondent gives his/her age whereby he/she appears younger or older (normally for prestige reasons). Errors in the reporting of the type of economic activity have also been found in censuses. Several other types of errors (memory lapse or recall lapse, and dating errors) in supplying information on many demographic variables have been found in censuses.

Further limitations to the use of census results in Africa are due to the slow processing, tabulation and publication of census material. Far too little attention has been paid to improvement in these phases of census operations. Many inconsistencies in the statistics and errors of calculation, and so on, are also common in African census results. Students should therefore note that all stages in the planning and execution of a census are potential sources of error and that close control must be exercised at all stages in order to minimise errors in census data.

Errors in Sample Survey Data

Two types of errors are found in sample surveys: sampling errors and non-sampling errors. The former is a measure of the degree to which the sample estimate differs from the true population value and occurs because a part, and not all, of the units is covered. This type of error is easy to control through the choice of suitable sample designs. The latter type of error, non-sampling error, is error other than sampling error and is common to both censuses and surveys. Sampling and non-sampling errors constitute what is known as total errors in sample surveys.

Errors that are common to both censuses and surveys can be classified under seven broad headings: (1) errors due to inadequate preparation, (2) errors of non-response, (3) observational errors, (4) processing errors, (5) errors in constructing estimates, (6) errors in interpretation, and (7) errors in publication. Non-sampling errors are likely to be greater in magnitude in censuses than in samples. Poor organisation and control, which are common features of surveys and censuses, particularly in developing countries, are the main cause. The errors that result from these can easily outweigh sampling errors for all practical purposes.

Examples of the errors and limitations enumerated under censuses and surveys are very common in Africa. Extensive reading of these reports prepared over a period give specific examples of the type of errors, inconsistencies in the data and warnings about their uses. A few examples from reports are used here for illustration.

(1) *Geographical Coverage and Distribution of Population*

An excellent example of the shortcomings in coverage and distribution of population in the 1967 census report of Tanzania is as follows:

> The one outstanding finding of the census, perhaps in one sense the most important, was the size of the population. Soon after the census, preliminary results gave a figure approximately one million above the old estimate. The discrepancy appears to be due mainly to poor coverage in the 1957 census, with consequent low estimates of population growth and would seem a strong indication of an almost complete coverage in 1967. Nevertheless, there are reasons to suspect that even the last census failed to reach the desired level in this respect. One study puts the maximum loss to as much as 2 per cent of the population.[2] Parts of the loss are due to almost insurmountable difficulties in the field work, given the conditions and limitations in manpower. Other parts can be referred to features in the organisation which in principle could have been avoided without too much cost, although the overall lack of

experience of the requirements of a census naturally led to a series of obstacles in optimum utilization of accessible resources. (Egero, 1973)

(2) *Absolute Population*

One main problem in population enumeration is how to record an individual, whether *de facto* (according to where he spent the night prior to the enumeration) or *de jure* (according to his usual place of residence). This problem is acute in Africa in view of frequent population movements. The experience of population censuses in Nigeria has found that:

> All de facto enumeration tends to under-estimate the population since the itinerant population as well as others temporarily away from home are omitted from the count. On the other hand, a de jure enumeration tends to over-estimate the people because it incorporates out-migrants who are permanently away from the community. (Arowolo, 1970)

Problems of interpretation of population statistics can also result when concepts used are left vague, as can be seen from the example from Ghana:

> While accounts of the 1891 and 1901 censuses do not firmly indicate the method in this regard, concepts used in the censuses suggest the probability of the counts being de jure. In the 1911, 1921 and 1931 censuses, however, the instructions clearly specified that the counts were to be de facto. In the 1948 census, a de facto count was intended, but the instructions in this regard were vague and non-specific. In the 1960 and 1970 censuses, the counts were primarily de facto, but de jure counts were primarily at the same time obtained by a breakdown of the population into 'usual members of a household' and 'visitors'. (Ghana, 1971)

An example of the general reliability of censuses in Africa comes from Uganda. In the introduction to the census returns of 1921, it was stated that:

> After examination of Mr Walter's covering report with the figures submitted, it was realised that while the principal totals in the tables relating to natives were, in the main, reasonable, there still existed so many inaccuracies and inconsistencies in the details as to render it impracticable to base any scientific analysis on the figures. (Kuczynski, 1948)

This opinion was used as an excuse for not publishing the report. It is quite possible that the data only showed the sort of inconsistencies that demographers in Africa have now learned to expect.

(3) *Sampling Procedure*

Official reports not uncommonly include references to failure to effect the sample design fully or to gain co-operation in all sampled areas. This statement from Tanzania gives an example from one African country:

> The sampling plan which is described in more detail in the second part of this chapter, has in principle fulfilled the requirements for a cluster sample. The enumeration areas (EA's) created in the geographic work were numbered serially within each division of the district. An enumeration area was then selected randomly among the first five, and the systematic selection of every fifth enumeration area gave a random sample of approximately 20 per cent of the population.
>
> Though simple in principle, the plan was difficult to put into effect partly through lack of adequate instructions and guidance . . . The net effects of deviations from the sample plan are very likely systematic errors or bias in the sample estimates. An over-representation of sparsely populated enumeration areas converted from non-typical to sample enumeration areas would be likely to have such consequences as would the non-random inclusion of special groups such as school children or hospital patients. (Egero, 1973)

(4) *Conceptual Problems*

Western concepts translated to the African environment can create serious difficulties in data collection. Tanzania's example is worth giving. In the census report of 1967, it was stated that:

> Local variations in the interpretation of concepts and methods were very common and may account for parts of the variation in results from the census. Economic data perhaps provide the best example. The economic classification of women apparently presented many difficulties due to their double role as a home-maker and agricultural worker and the instructions offered little guidance in their heavy reliance on conditions in an industrialized type of society. The solutions chosen also varied strongly; in some districts the vast majority, in others only a small proportion of the women were classified as economically active. The concept (gainfully) 'employed' similarly caused problems in interpretation. In Mafia district only persons employed for wages were classified in this

category and peasants and fishermen were regarded as 'others not working'. All sample EA's in this district had to be re-enumerated to correct for this error. (Egero, 1973)

(5) *Age Data*

The censuses conducted in the majority of African countries have generally asked for the age of the individual enumerated either in completed years or in months.[3] The information is useful for almost every aspect of development planning – for studies of employment, manpower and the economic structure of the population – for educational planning, and for the analysis of fertility and population projections. Unfortunately, in Africa the accurate estimation of ages constitutes one of the most difficult problems that face the census administration, simply because of the widespread ignorance of age in the orthodox Western sense of the completed number of years or months. The formal question is either on the concept of 'age at last birthday' or 'age at next birthday' or 'age at nearest birthday'. This is a typical question with its implied Westernised concepts of celebration of birthdays.

There are at least two possibilities of error in this type of question. First, in most countries of Africa, where the population is predominantly rural, people do not remember their age by celebrating their birthdays. Second, people just do not know their birthdays.[4] The use of the well-known technique of historical calendars to assist enumeration in pinpointing the dates of birth of respondents has produced some bizarre results.[5]

The net effect of these problems is that the age data are not very reliable in African countries. The problem is well summarised by this example from Tanzania:

> Of the demographic variables, none could be said to be completely free from errors. Evidence on under-reporting points to sex selectivity, young men being the most difficult category to cover. Age distortions are heavy, a combination of digital preference and systematic over-estimation when the use of an events calendar with about ten events only added further to the otherwise fairly regular age heaping pattern. Under-reporting of young children seems also to have affected the distribution. (Egero, 1973)

(6) *Population Growth*

The growth of a population is a function both of its natural increase (births minus deaths) and of migration. Growth estimates based on comparisons between two or more consecutive censuses take care of all these factors and are therefore reliable, assuming, of course, that the census figures are unbiased. More often than not there is underestimation and/or overestimation in censuses in Africa, as the example of Tanzania shows:

> Observed growth rates for the inter-censal periods are 1·8 per cent for 1948 – 57 and 3·1 per cent for 1957 – 67, a difference which if accepted would imply an unrealistically high decline in the death rate . . . [Moreover] a look at the observed growth rates per district in the two inter-censal periods discloses that, although most of them are higher in the period 1957 – 67, for about one-third of the districts they are more than three times as high as in the period 1948 – 57. Such differences are in most cases difficult to explain in terms of changing patterns of fertility, mortality or migration. (Gunnar, 1968)

Thus, a comparison between two census figures of which the first one is underestimated (or more underestimated than the second) gives an exaggerated growth estimate.

(7) *Fertility and Mortality*
It is no exaggeration to claim that statistics on fertility and mortality that are based on questions on births and deaths in the household are not only unreliable but even useless in any form of planning unless adjusted in one way or another. This view has been clearly put by Blacker (1971) as follows:

> Such data are, it is well known, subject to serious errors. Anyone who accepts the results of these (retrospective) questions at their face value without first subjecting the data to stringent tests, can only be regarded by his fellow demographers as both naive and ignorant.

The following two examples from Nigeria and Tanzania show that the warning by Blacker should be taken seriously by all students of African statistics. Commenting on the results of the 1965 – 6 demographic surveys, the following statement from Nigeria mirrors the problems faced by most African countries in the analysis of fertility and mortality statistics:

> It was found that the entire material could not be accepted for analysis. In some sample units, the recorded birth rate was as low as 1 – 5 per 1,000 and the death rate as low as 2 – 4 per 1,000. It was clear that in such cases considerable non sampling errors accounted for the deflated errors. (Omoh, 1970)

From Tanzania, the following statement echoes the same sentiment on the unreliability of fertility and mortality statistics and the general problems or difficulties of collecting data in Africa:

> The question on current fertility appears to have resulted in over-reporting of children born due to misconceptions about the reference period. Retrospective fertility on the other hand, would seem to be under-reported, a common error due to memory lapse and exclusion of children deceased soon after birth. One question was, optimistically, directly inquiring into the number of persons deceased in the household during the last year; for many, the best way to evade the question appears to have been a simple denial of any deaths. (Egero, 1973)

The reluctance encountered in Tanzania to tell enumerators about deceased children is widespread in Africa. Many of the children born alive, but who later were deceased, would never be mentioned; consequently, the number of births might also be under-reported. Another source of error in death-reporting is when a person who has lived alone dies. This death is more likely to be missed or not reported because the house may be deserted and there may be nobody to report this death. Errors in age-reporting further complicate the analysis of measures of fertility and mortality. There is a range of models to adjust for these errors, but virtually none of them can be claimed to give unbiased estimates of basic vital rates.

Errors in Vital Registration Data

Basically, the evaluation of the reliability of vital statistics involves measuring both the qualitative and the quantitative accuracy with which vital events are either registered or reported in a survey, and the characteristics of the events that are reported by the informant and recorded by the registrar. As in other forms of demographic data collection, the quality of data from vital registration systems is influenced by methods adopted in their collection and organisation. They are therefore subject to various kinds of systematic errors. Some of these types of error seem to be almost universal, while other patterns of distortion are limited to specific cultural or administrative conditions: for example, a tendency to report or register infants of one sex more completely than those of the other.

In general, the patterns of age mis-statement of the mother or the deceased are similar to those found in census data. Specifically in the case of deaths, vital registration data are subject to even greater inaccuracy than census data, because the information on age at death in cases where the birth certificate is not available is obviously obtained from someone other than the person who has died. Digital preference is therefore observed to a greater extent on death certificates than in census data (Linder, 1959).

The completeness of registration also varies considerably by type of event; for example, infants who have died may be under-reported or missed in both the birth and the death registers. The accessibility of the place of occurrence also influences completeness of registration. Births and deaths that take place in rural areas, especially when these areas are remote, or are inaccessible, are less likely to be registered. Various factors relating to the method of operation of the registration system also affect the completeness of registration; for example, registration is less likely to be complete if the responsibility for registration is vague, or undefined, than if specific individuals are required to register the events of birth and death. Specific examples of the various types of errors can be cited from a number of African countries where the system is operating. Examples of these errors are as follows.

(1) *Coverage of Total Events*

Reference has already been made to the low quality of demographic information available in Africa, and the field of vital registration data is no exception. Examples from the existing system show that there is a very substantial under-registration of events as well as various inaccuracies in the recording of events that are registered. This example from Ghana states:

> In terms of population, the registration districts in Ghana in which registration offices have been established, form about 95 per cent of the entire population. With a crude birth rate of 47 live births per 1,000 population, the expected total births and deaths from these registration areas are 382,000 births and 170,000 deaths respectively. In 1970 only 85,000 live births and 28,000 deaths were registered; this is only about 22 per cent of the expected births and 17 per cent of the expected deaths. (Owusu, 1972)

The low coverage of births and deaths in the registration districts in Ghana in recent years is due, no doubt, to the rapid expansion of registration areas without much attention being paid to the quality of the data. It can be attributed largely also to the inadequacy of the existing local registration offices by which registration facilities are brought nearer to the people, and the inability of the offices to establish an adequate number of local registration offices is also attributed to lack of funds to employ an adequate number of full time staff to man these services.

In Kenya, the 'Report of the Registrar of Births and Deaths' in the

Statistical Abstract (1971) remarks that the statistics collected may be considered reliable in respect of Europeans and Asians but that, for Africans, births and deaths are under-registered. It further notes that the recent increase in births and deaths registered in each area is partly a reflection of improved coverage and increased public awareness of the advantages to be derived from civil registration.

From the annual return of vital statistics (Katsina Province) in Nigeria in 1966, it was obvious that the indicated frequencies of births and deaths in the population were far below those to be expected; for example, the total population in 1963 for the whole province was 2,545,006, with total births of 65,502, giving a birth rate of about 25·00 per 1,000. The same comments apply to the death rates and infant mortality rates, where the statistics declared show a death rate of 8·5 per 1,000 and an infant mortality of 72·9 per 1,000 live births. Umoh (1972) has noted that the existing registration ordinance does not make provision for a central or local agency that, among other things, could gather, analyse and disseminate basic vital statistics. The result is that, although births and deaths are being registered in towns all over Nigeria, records of such registration are not available to the Federal Office of Statistics.

Several other studies (e.g. Zaghloul and El-Ghamry, 1971) of registration coverage undertaken in African countries tell essentially the same story. For example, in Tunisia, the rate of under-registration of births varied between 7 and 11 per cent in the period 1961 – 9; in Algeria, the percentage of under-registration between 1963 and 1967 was about 7 per cent; whereas in Libya, during the period 1964 – 7, the rate of under-registration was about 30 per cent; the comparative figure for Egypt for the period 1960 – 5 was about 5 per cent.

(2) *Late Reporting of Events*

In Uganda, the published statistics for Asians and Europeans seem to be reasonably accurate, but there is some evidence to show that the proportion of events that are recorded as late registration is not negligible. For example, in 1958, 78·7 per cent of all Asian births registered occurred in 1958, 15·5 per cent occurred in 1957, and 5·8 per cent occurred prior to 1957. With deaths, 88 per cent occurred in 1957 and only 1·8 per cent in other years. In Ghana, where the same type of analysis was carried out: 'The pattern of late registration was much higher for births than for deaths' (Kpedekpo, 1970). Registration systems have the notable advantage of being able to record information about events soon after they occur, and such promptness serves to increase the completeness of the registration system, as well as the quality of the information obtained about items included on the registration forms. The fact that some substantial late registration occurs in Uganda and Ghana may throw some doubt on the accuracy and completeness of the published statistics.

In all African countries where statistical compilations are available from registration data, considerable uncertainty exists about the actual completeness of many national registration systems; in particular, erroneous conclusions can easily be reached about levels and trends in the vital rates because many countries with improving registration systems fail to conduct tests of registration completeness. One serious limitation to the use of the data from such vital registration systems as far as coverage is concerned is that the degree of under-registration, and the extent to which vital events are omitted from the register, limit its use as a source for calculating national birth and death rates. Inaccuracies in the data compiled also limit their usefulness as a means of learning the facts of the country's social, demographic and health situation. This example from Tanzania emphasises the point:

> The registration of births and deaths and marriages has been compulsory for Europeans and Asians since 1921. In 1966, the ordinance was extended to include Africans in all townships. The registration is incomplete in nature and therefore of no statistical value. (Tanzania, 1971a)

CONCLUDING REMARKS

We have already drawn attention to the important role that population data play in the analysis of social and economic phenomena. The availability of demographic data in sufficient quantity and quality is therefore a necessary prerequisite for the analysis and understanding of many aspects of social and economic analysis.

NOTES: CHAPTER 2

1 The cost of the 1973 Nigerian census was 37 million Naira or about US $59·2 million (see, for example, Udo, 1978).

2 See, for example, Tanzania (1971b).

3 In the 1959 census of Uganda, for example, the following age groups were adopted: under 1, 1 – 5, 6 – 15, 16 – 45, over 45 years. The 1952 – 3 Nigerian census adopted the following groupings: under 2, 2 – 6, 7 – 14, 15 – 49, 50 and over years. Such an age-grouping, although useful in the Nigerian and Ugandan situations, fails to conform to international standards.

4 The only countries where the question on age has been replaced by the date of birth are Rhodesia (1962, 1969), Zambia (1963, 1969) and Malawi (1966). In Rhodesia (1962) and Zambia (1963), the question did not require the date of birth by individual years for the whole population; persons over 16 years in Rhodesia and

over 21 years in Zambia were merely allocated to broad age groups.

5 See, for example, the case study on Ghana by de Graft-Johnson and Ramachandran (1973).

REFERENCES AND FURTHER READING

Arowolo, O. O. (1970), *The Problem of Population Movement and Enumeration in Nigeria: The Example of Ife Division of the Western State* (Philadelphia: Population Studies Center, University of Pennsylvania).

Blacker, J. G. C. (1971), 'The estimation of vital rates from census data in Kenya and Uganda', paper presented at the African Population Conference, Accra (9 – 18 December).

Bondestam, L. (1973), *Some Notes on African Statistics: Collection, Reliability and Interpretation*, Research Report No. 18 (Uppsala: Scandinavian Institute of African Studies).

Brass, W. (1968), 'The improvement in the quantity and quality of demographic statistics', in Caldwell and Okonjo (1968).

Caldwell, J. C. and Okonjo, C. (eds) (1968), *The Population of Tropical Africa* (London: Longman).

Conde, J. (1971), 'Non-conventional sources of demographic data', paper presented at the African Population Conference, Accra (9 – 18 December).

De Graft-Johnson, K. T. and Ramachandran, K. V. (1973), 'Evaluation and analysis of the 1970 population census results of Ghana', paper presented at a conference organised by the Economic Commission for Africa, Accra (16 – 28 July).

Des Raj (1972), *The Design of Sample Surveys* (New York: McGraw-Hill).

Economic Commission for Africa (annual), *Demographic Handbook for Africa* (Addis Ababa).

Egero, B. (1973), 'Errors in the census of the population of Tanzania', Census Vol. 6, in Egero and Henin (1973).

Egero, B. and Henin, R. A. (eds) (1973), *The Population of Tanzania: An Analysis of the 1967 Population Census*, Census Vol. 6 (Dar es Salaam: Bureau of Statistics).

Ghana (1971), Country statement prepared for the African Population Conference, Accra (9 – 18 December), No. POP/INP/79, pp.11-12.

Gil, B. and Omaboe, E. N. (1965), 'Population censuses and national sample surveys in developing countries', *International Labour Review*, vol. 92, pp.169 – 83.

Gunnar, J. (1968), *Provisional Estimates of Fertility, Mortality and Population Growth for Tanzania* (Dar es Salaam: Central Bureau of Statistics, Ministry of Economic Affairs and Development Planning), table 12.

Kenya (1971), *Statistical Abstract* (Nairobi: Ministry of Economic Planning).

Kpedekpo, G. M. K. (1967), 'The nature and extent of errors in the 1960 Ghana population census', *West African Journal of Science*, vol. 12, pp. 149 – 63.

Kpedekpo, G. M. K. (1968), 'A note on the errors in the age data in the 1960 census', *West African Journal of Science*, vol. 13, pp. 127 – 34.

Kpedekpo, G. M. K. (1968), 'Evaluation and adjustment of vital registration data for the compulsory registration areas of Ghana', *American Journal of Demography*, vol. 5, no. 1, pp. 86–92.

Kpedekpo, G. M. K. (1970), *Studies on Vital Registration Data from Compulsory Registration Areas of Ghana*. Vol. 2: *An Analysis of Seasonality of Births and Deaths* (Legon, Accra: Institute of Statistical, Social and Economic Research, University of Ghana).

Kpedekpo, G. M. K. (1971), 'Patterns of delayed registration of births and deaths in Ghana', *Ghana Social Science Journal*, vol. 1, pp. 1–20.

Kpedekpo, G. M. K. (1972), 'Some indicators of age mis-statement in the 1969 Uganda census', *West African Journal of Science*, vol. 17, no. 2, pp. 71–84.

Kpedekpo, G. M. K. (1979), *Essentials of Demographic Analysis for Africa* (London: Heinemann), ch. 1.

Kpedekpo, G. M. K. and Chacko, V. J. (1969), 'Sampling for demographic characteristics in Ghana', *Sankhya: A Journal of the Indian Statistical Association*, series B, vol. 30.

Kuczynski, R. R. (1948), *Demographic Survey of the British Colonial Empire*, Vol. 1 (London: Oxford University Press).

Kuczynski, R. R. (1949), *Demographic Survey of the British Colonial Empire*, Vol. 2, (London: Oxford University Press), ch. 9.

Lury, D. A. (1968), 'Population data of East Africa', in Caldwell and Okonjo (1968).

Office of Population Research at Princeton University and Population Association of America Inc. (quarterly), *Population Index*, a bibliography (Princeton, NJ, joint issue).

Omoh, O. E. (1970), Statement by Nigeria's participant at the Working Group on Fertility Studies and Evaluation of Population Programmes (26–30 January), No. POP/INF/32.

Owusu, J. Y. (1972), Report on registration in Ghana, Office of the Government Statistician, Accra (unpublished).

Tanzania (1971a), 'Demographic information in Tanzania', paper prepared for the African Population Conference, Accra (9–18 December).

Tanzania (1971b), *Census Methodology*, Census Vol. 5 (Dar es Salaam: Bureau of Statistics).

Udo, R. K. (1968), 'Population and politics in Nigeria', in Caldwell and Okonjo (1968).

Udo, R. K. (1978), 'Thinking of a census before 1983', *Daily Times of Nigeria* (13 May).

Uganda (1959), *Annual Report of the Administrator General, 1958–59* (Kampala: Ministry of Justice).

Umoh, O. E. (1972), 'Demographic statistics in Nigeria', in S. H. Ominde and C. N. Ejiogu (eds), *Population Growth and Economic Development in Africa* (London: Heinemann).

United Nations (1965), *Handbook on Population Census Methods*. Vol. 1: *General Aspects of a Population Census*. Vol. 2: *Economic Characteristics of the Population*. Vol. 3: *Demographic and Social Characteristics of the Population* (New York: UN Statistical Office).

United Nations (annual), *Demographic Yearbook* (New York: UN Statistical

Office, Department of Economic and Social Affairs).
van de Walle, E. (1968), 'Characteristics of African demographic data', in W. Brass *et al.* (eds), *The Demography of Tropical Africa* (Princeton, NJ: Princeton University Press), ch. 2.
Wogugu, M. O. (1974), 'Demographic data in sub-Saharan Africa: an evaluation of their adequacy for basic research', paper presented at a meeting of the Population Association of America Inc., New York.
Zaghloul, S. and El-Ghamry, A. (1971), 'Present levels and trends of fertility in Arab countries', in *The Fertility Trends and Differentials in Arab Countries*, Research Monograph Series 2 (Cairo: Cairo Demographic Centre).

MAJOR SOURCES OF POPULATION STATISTICS

Abbreviations:
INSEE: Institut National de la Statistique et des Etudes Economiques.
CICRED: Committee for International Co-ordination of Research in Demography.

ALGERIA
Negadi, G., Tabutin, D. and Vallin, J., 'Situation démographique de l'Algérie', in *La Population de l'Algérie*, national population monograph in the CICRED series (Paris, France, 1974), ch. 2.

ANGOLA
Repartição de Estatistics Geral, *3° Recenseamento Geral de População, 1960*, Vol. 1 (Luanda, 1964).
Paxton, J. (ed.), *The Statesman's Year-Book* (London, UK: Macmillan, 1977).
Instituto Nacional de Estatística (Portugal), *Anuário Estatístico Territórios Ultramarinos*, Vol. 2 (Lisbon, Portugal, 1975).

BENIN (formerly DAHOMEY)
Ministère de la Coopération (Benin) and INSEE, Service de Coopération (France), *Enquête démographique au Dahomey, 1961: résultats definitifs* (Paris, France, 1964).
Survey population excludes Abomey-Bohicon and Tchi canton (1 per cent of population).

BOTSWANA
Central Statistics Office, *Report of the Population Census, 1971* (Gaborone, 1972).
Central Statistics Office, *Statistical Abstract, 1972* (Gaborone, 1972).

BURUNDI
Département de Statistiques, *Annuaire Statistique, 1968* (Bujumbura, 1970).
INSEE (France), *Enquête démographique, 1970–1971*. Vol. 1: *Résultats definitifs*. Vol. 2: *Methodologie de l'enquête* (Paris, France, 1974).

CAMEROUN
Service de la Statistique (Cameroun) and Ministère de la Coopération, INSEE (France), *Enquête démographique au Cameroun, 1962–64: résultats definitifs pour la région nord* (Paris, France, 1968).
Service de la Statistique (Cameroun) and Ministère de la Coopération, INSEE (France), *Enquête démographique au Cameroun, 1962–64: résultats definitifs pour la région sud est* (Paris, France, 1968).
Service de la Statistique (Cameroun) and Société d'Etudes pour le Développement Economique et Social (France), *The Population of West Cameroun: Main Findings of the 1964 Demographic Sample Survey* (Paris, France, 1966).

North Cameroun survey population excludes nomadic M'bororos (15,000) and Tikar plain area (4,000). Southeast Cameroun survey population excludes Yaounde and Douala (280,000) and Bamiléké area (1,120,000).

Direction de la Statistique (DS, Cameroun) and Société d'Etudes pour la Développement Economique et Social (France), *La Population du pays Bamiléké et des départements limitrophes: principaux résultats de l'enquête démographique de 1965* (Paris, France, 1966).

Direction de la Statistique et de la Comptabilité Nationale (DSCN), *Principaux Résultats du recensement général de la population et de l'habitat d'avril 1976* (Yaounde, 1978).

Service de la Statistique et de la Mécanographie (SSM), *Recensement de la ville de Yaounde en 1962: principaux résultats* (Yaounde, 1963).

Service de la Statistique Générale (SSG, Cameroun) and INSEE, Service de Coopération (France), *Enquête démographique au Cameroun: résultats definitifs pour la région nord, 1962–1964* (Paris, France, 1968).

Direction de la Statistique (DS, Cameroun) and Société d'Etudes pour le Développement Economique et Social (France), *La Population du Cameroun occidental: résultat de l'enquête démographique de 1964* (Paris, France, 1969).

CENTRAL AFRICAN REPUBLIC

Service de la Statistique Générale (Central African Republic) and Ministère de la Coopération, INSEE (France), *Enquête démographique en Republique Centrafricaine, 1959-60: résultats de finitifs* (Paris, France, 1968).

Survey population excludes eastern portion of country (66,000), nomadic groups (40,000) and Bangui town (80,000).

INSEE (France), *Afrique Noire, Madagascar, Comores: démographie comparée*, Vol. 1 (Paris, France, 1967).

Secrétariat d'Etat aux Relations avec les Etats de la Communauté (France), *Données de base sur la situation démographique* (Paris, France, 1961).

CHAD

Service de Statistique (Chad) and INSEE, Service de Coopération (France), *Enquête démographique au Tchad, 1964: résultats definitifs*. Vol. 1: *Analyse des résultats*. Vol. 2: *Tableaus statistiques detailles* (Paris, France, 1966).

CONGO

Service de Statistique (Congo) and INSEE, Service de Coopération (France), *Enquête démographique, 1960–61: résultats definitifs* (Paris, France, 1965).

Survey population excludes Brazzaville (136,000) and Pointe-Noire (79,000).

INSEE (France), *Afrique Noire, Madagascar, Comores: démographie comparée* (Paris, France, 1967).

ETHIOPIA

Central Statistical Office, *Results of the National Sample Survey Second Round*. Vol. 1: *The Demography of Ethiopia*, Statistical Bulletin 10 (Addis Ababa, 1974).

Central Statistical Office, *Statistical Abstract of Ethiopia, 1975* (Addis Ababa, 1975).

Central Statistical Office, *Statistical Abstract, 1971* (Addis Ababa, undated).

GABON

Service National de la Statistique (SNS), *Recensement de la population, 1960–61*. Vol. 1: *Données d'ensemble* (Libreville, 1964).

Service de Statistique (SS, Gabon) and INSEE, Service de Coopération (France), *Recensement et enquête démographique, 1960–61: résultats definitifs: ensemble du Gabon* (Paris, France, 1965).

GAMBIA

Oliver, H. A., *Report of the Census of Population Taken on 17th–18th April, 1963* (Bathurst, 1965).

Central Statistics Division, *Population Census 1973: Statistics for Local*

Government Areas and Districts, Vol. 3 (Banjul, 1976).

GHANA

Census Office, '1970 population census of Ghana: the 1971 supplementary enquiry', unpublished, Accra, 1975.

Census Office, *1970 Population Census of Ghana*. Vol. 3: *Detailed Demographic Characteristics* (Accra, 1975).

Gaisie, S. K., *Dynamics of Population Growth in Ghana* (Accra-Tema: Ghana Publishing Corporation, 1969).

GUINEA

Guinea and Ministère de la France d'Outre-Mer, Service de Statistiques (France), *Etude démographique par sondage en Guinee, 1954–55: résultats definitifs* (Paris, France, undated).

Conde, J., 'La Situation démographique en République de Guinee', *Revue française d'études politiques africaines*, no. 123 (Paris, France, March 1976).

GUINEA (BISSAU)

Provincia da Guiné (Portugal), *Censo da População de 1950*. Vol. 2: *População não Civilizada* (Lisbon, Portugal, undated).
Data for the 'non-assimilated' African population only.

IVORY COAST

Service de la Statistique (Ivory Coast) and Ministère de la France d'Outre-Mer, Service de Statistiques (France), *Etude démographique du 1er secteur agricole de la Côte d'Ivoire, 1957–58: résultats provisoires* (Paris, France, 1958).

Ministère du Plan, *Côte d'Ivoire, 1965, population: études régionales, 1962–1965, synthèse* (Abidjan, 1967).

KENYA

Ministry of Finance and Economic Planning, *Kenya Population Census*, Vol. 1 (Nairobi, 1970).

Ministry of Finance and Economic Planning, *Kenya Population Census, 1969*. Vol. 2: *Data on Urban Population* (Nairobi, 1971).

Ministry of Finance and Economic Planning, *Kenya Statistical Digest*, vol. 4, no. 2 (Nairobi, 1971).

LESOTHO

Bureau of Statistics, *1966 Population Census Report*, Vol. 1 (Maseru, undated).

LIBERIA

Department of Planning and Economic Affairs, *Demographic Annual of the Population Growth Survey, 1970: Patterns of Natality*, Series N-1 (Monrovia, 1971).

Ministry of Planning and Economic Affairs, *Demographic Annual of the Population Growth Survey, 1970: Demographic Patterns in Liberia, 1970*, Series S-1 (Monrovia, 1972).

Ministry of Planning and Economic Affairs, *Demographic Annual of the Population Growth Survey, 1971: Natality Profiles*, Series N-2 (Monrovia, 1972).

Ministry of Planning and Economic Affairs, *Demographic Annual of the Population Growth Survey, 1971: Demographic Profiles in Liberia, 1971*, Series S-2 (Monrovia, 1973).

Ministry of Planning and Economic Affairs, *1974 Census of Population and Housing: Final Population Totals and Related Percentages with Some Salient Demographic Characteristics*, Population Bulletin no. 2 (Monrovia, 1976).

Office of National Planning, Bureau of Statistics, *1962 Population Census of Liberia* (Monrovia, 1965).

LIBYA

Census and Statistical Department, *Population Census (Preliminary Results)* (Tripoli, 1973).

Census and Statistical Department, *Statistical Abstract of Libya, 1971* (Tripoli, undated).

Census and Statistical Department, *Vital Statistics, 1972* (Tripoli, undated).

MADAGASCAR

Institut National de la Statistique et de la Recherche Economique (INSRE), *Enquête démographique Madagascar, 1966* (Tananarive, 1967).

MALAWI

National Statistical Office, *Malawi Population Census, 1966: Final Report* (Zomba, 1969).

National Statistical Office, *Malawi Population Change Survey, February 1970 – January 1972* (Zomba, 1973).

MALI

Service de la Statistique (Mali) and INSEE, Service de Coopération (France), *Enquête démographique au Mali, 1960–61* (Paris, France, undated).

Mission Socio-économique 1956–1958 (Mali) and Ministère de la Coopération, INSEE (France), *Enquête démographique dans le delta central nigerien*, Vols 1 and 2 (Paris, France, undated).

MAURITANIA

Service de la Statistique (Mauritania) and Société d'Etudes pour le Développement Economique et Social (France), *Enquête démographique, 1964–65: résultats provisoires* (Paris, 1966).

Survey population excludes nomad subdivisions of Tichitt and Boghe (20,000) and urban centres (70,000).

Service de la Statistique (Mauritania) and INSEE, Département de Coopération (France), *Enquête démographique, 1965: résultats definitifs* (Paris, 1972).

MAURITIUS

Central Statistical Office, *1972 Population Census of Mauritius*. Vol. 1: *Preliminary Report* (Rose Hill, 1974).

Central Statistical Office, *Bi-Annual Digest of Statistics, December 1975*, Vol. 10, No. 2 (Rose Hill, 1976).

MOROCCO

Centre de Recherches et d'Etudes Démographiques (CERED), Direction de la Statistique, *La Fécondité marocaine*, No. 6 (Rabat, 1974).

Direction de la Statistique, *Tables de mortalité marocaine*, (Rabat, 1974).

Direction de la Statistique, *Recensement général de la population et de l'habitat, 1971: résultats du sondage au 1/10* (Rabat 1972).

Direction de la Statistique, *Recensement général de la population et de l'habitat, 1971: population active: résultats du sondage au 1/10* (Rabat, 1973).

Institut National de Statistique et d'Economie Appliqué (INSEA), *La Population du Maroc*, national population monograph in the CICRED series (Rabat, 1974).

MOZAMBIQUE

Instituto Nacional de Estatística, *IV Recenseamento Geral da População, 1970: Resumo Geral* (Lourenço Marques, 1974).

NIGER

Mission Démographique de Niger, 1960 (Niger) and Ministère de la Coopération, INSEE (France), *Etude démographique de Niger: résultats definitifs* (Paris, France, 1963).

Service de la Statistique et de la Mécanographie, *Annuaire Statistique* (Paris, France, 1967).

NIGERIA

Acsadi, G. T., Igun, A. A. and Johnson, G. Z., *Surveys of Fertility, Family and Family Planning in Nigeria*, Publication No. 2, (Ile-Ife: Institute of Population and Manpower Studies, University of Ife, 1972).

Federal Office of Statistics, *Rural Demographic Sample Survey, 1965–1966*, (Lagos, 1968).

Igun, A. A. and Acsadi, G. T. (eds), *Demographic Statistics in Nigeria*, proceedings of a symposium (Ile-Ife: Demographic Research and Training Unit, University of Ife, 1972).

Iro, M. I., 'The demography of Nigeria, 1950–66: with special reference to the methods and accuracy of the population censuses during this period', Ph.D. dissertation, Cornell University, Ithaca, NY (1973).

Mott, F. L. and Fapohunda, O. J., *The Population of Nigeria*, Monograph No. 3 (Lagos: Human Resources Unit, University of Lagos, 1975).

Department of Statistics, *Population Census of the Eastern Region of Nigeria, 1953* (Lagos, undated).

Department of Statistics, *Population Census of the Northern Region of Nigeria, 1952* (Lagos, undated).

Department of Statistics, *Population Census of the Western Region of Nigeria, 1952* (Lagos, undated).

Federal Office of Statistics, *Rural Demographic Survey, 1965–66*, (Lagos, 1968).

RWANDA

Office Général de Statistiques, *Enquête démographique, 1970*, Vols 1 and 2 (Paris, 1973).

SENEGAL

Direction de la Statistique, *Enquête démographique nationale 1970–71: analyse des résultats du 2ème passage: portant sur la population active* (Dakar, 1973).

Direction de la Statistique, *Enquête démographique nationale 1970–71: résultats definitifs*. Vol. 1: *Pyramids des ages* (Dakar, 1974).

Direction de la Statistique, *Résultats provisoires du recensement général de la population d'avril 1976* (Dakar, undated).

SEYCHELLES

Census Commissioner, *1977 Census Report* (Victoria, 1978).

SIERRA LEONE

Central Statistics Office, *1963 Population Census of Sierra Leone*. Vol. 1: *Number of Inhabitants* (Freetown, 1965).

Central Statistics Office, *Quarterly Statistical Bulletin*, Vol. 2 (Freetown, June 1966).

SOMALIA

Statistical Department, *Somalia Statistics* (Mogadishu, 1969).

SUDAN

Demeny, P., 'The demography of the Sudan: an analysis of the 1955–56 census', in W. Brass *et al.* (eds), *The Demography of Tropical Africa* (Princeton, New Jersey, USA: Princeton University Press, 1968).

Ministry of Finance, Planning and National Economy, *Economic Survey, 1975–76* (Khartoum, 1976).

National Planning Commission, *Economic Survey, 1974* (Khartoum, 1975).

SWAZILAND

Jones, H. M., *Report on the 1966 Swaziland Population Census* (Mbabane: Census Office, 1968).

Department of Statistics, *Swaziland Annual Statistical Bulletin, 1970* (Mbabane, 1970).

TANZANIA

Bureau of Statistics, Ministry of Economic Affairs and Development Planning, *1967 Population Census*, Vols 1–6 (Dar es Salaam, 1973).

Bureau of Statistics and University of Dar es Salaam, Bureau of Resource Assessment and Land Use Planning, *The Demography of Tanzania*. Vol. 6: *An Analysis of the 1973 National Demographic Survey of Tanzania* (New York, USA, undated).

Bureau of Statistics and University of Dar es Salaam, Bureau of Resource Assessment and Land Use Planning, *1973 National Demographic Survey of Tanzania*. Vol. 1: *Regional and National Data*. Vol. 4: *The Methods Report* (Dar es Salaam, undated).

Central Statistical Bureau, *Provisional Estimates of Fertility, Mortality and Population Growth for Tanzania* (Dar es Salaam, 1968).

TOGO

Direction de la Statistique, *Recensement général de la population, mars–avril 1970*, Vol. 1 (Lomé, 1974).

Service de la Statistique Générale, *Enquête démographique, 1961*. Vol. 2: *résultats definitifs* (Lomé, undated).

TUNISIA

Institut National de la Statistique, *Recensement général de la population et des logements, 8 mai 1975*. Vol. 1: *Logements, tableaux et analyses des résultats de sondage au 1/10 ème* (Tunis, undated).

Institut National de la Statistique, *Recensement général de la population et des logements, 8 mai 1975: population par division administrative* (Tunis, undated).

UGANDA

East African Statistical Department, *Uganda Census, 1959: Non-African Population* (1960).

East African Statistical Department, *Uganda Census, 1959: African Population* (1961).

Ministry of Finance, Planning and Economic Development, *Report on 1969 Population Census: Additional Tables*, Vols 1–3 (Entebbe, 1973).

Ministry of Finance, Planning and Economic Development, *1971 Statistical Abstract* (Entebbe, undated).

UPPER VOLTA

Courel, A. and Pool, D. I., 'Upper Volta', in J. C. Caldwell (ed.), *Population Growth and Socioeconomic Change in West Africa* (New York, USA: Columbia University Press, 1975).

Service de Statistique (Upper Volta) and Ministère de la Coopération, INSEE (France), *La situation démographique en Haute-Volta: résultats partiels de l'enquête démographique, 1960–61* (Paris, France, 1962).
Survey population excludes Ouagadougou (60,000) and Bobo-Dioulasso (50,000).

Service de Statistique et de la Mécanographie (Upper Volta) and INSEE (France), *Enquête démographique par sondage en République de Haute-Volta, 1960-1961*, Vols 1 and 2 (Paris, France, 1970).

ZAIRE

Bureau de la Démographie, Tableau Général de la Démographie Congolaise, *Enquête démographique par sondage, 1955–57: analyse générale des résultats statistiques* (Leopoldville, 1961).
Information by districts:
(1) Congo Belge, 2e Direction Générale, le Direction, AIMO, *Enquêtes démographiques*, Vols 1–5 (Cîté Leopoldville, Territoire Suburbaine, Bas-Congo, Cataractes, Tshuapa) and Vol. 7 (Maniema).
(2) Congo Belge, Affaires Economiques, Direction de la Statistique, *Bulletin mensuel de statistiques générales du Congo Belge et du Ruanda-Urundi*, Special Series No. 3, *Enquête démographique, 1956–57*, Vols B, E and F (Equateur Province excluding Tshuapa District, Katanta Province and Kasai Province).
(3) République du Congo, Ministère du Plan et de la Coordination Economique, Service des Statistiques, *Résultats de l'enquête démographique*, Vols 11–13 (Lac Leopold II, Kwilu, Kwango, Nord- and Sud-Kivu, Katanga).
Survey population excludes part of Katanga (52,000).

Bureau de Président, Planification du Développement, *Perspectives démographiques provisoires pour la République du Zaire, 1970–1980*, with the collaboration of the Institut National de la Statistique, Kinshasa.

ZAMBIA

Central Statistical Office, *Census of Population and Housing, 1969: First Report* (Lusaka, 1970).
Central Statistical Office, *Census of Population and Housing, 1969: Final Report*. Vol. 1: *Total Zambia* (Lusaka, 1973). Vol. 3: *Demographic Analysis* (Lusaka, 1974).

Central Statistical Office, *Sample Census of Population, 1974: Preliminary Report* (Lusaka, 1975).

Ohadike, P. O. and Tesfaghiorghis, H., *The Population of Zambia*, National population monograph in the CICRED Series (Paris, France, 1975).

CHAPTER 3

Health Statistics

We must start this chapter by drawing attention to the important and growing connection between health-planning and the need for health statistics. A number of planning methods have been used in recent years in an attempt to integrate the health sector into a general development plan for most countries. These range from empirical methods, in which priorities are largely based on what is felt to be the magnitude of the problem, to methods involving mathematical simulation, in which the health sector is treated like an industry with inputs leading to outputs. The methods involve a great deal of difficult calculations, which should normally be based on reliable information. Unfortunately, African countries have not got the type of information or data required for this exercise.

Nevertheless, the increasing desire of governments to improve health conditions and to engage in health-planning has lead to the need to search for better data on all aspects of health. But, as we have clearly noted, the kinds of health statistics required are dependent on the type of health-planning carried out in these countries. The health condition of a population is also reflected in its mortality and morbidity; this in turn affects fecundity and fertility; therefore, most data on health deal with mortality, morbidity and fertility. Furthermore, mortality and morbidity are influenced by the processes of economic development, modernisation, medical technology and improvements in government organisation, distribution of health services, and public health legislation and practice; therefore, data on some of these aspects of health-planning may also be collected.

Data on health required for planning can be grouped under five main headings: (1) population data; (2) health problems data, mainly on morbidity and mortality; (3) health services resources (input data); (4) health services achievements (output data); and finally, (5) health services costs. This chapter treats sources of health statistics, the types of data collected, their uses, limitations or weaknesses, and problems of interpretation.

SOURCES OF DATA

Data on various aspects of a country's health can be obtained from (1) vital registration systems (births, marriages and deaths), (2) hospital records, (3) records of health authorities or health ministries, (4) general or private practitioners' records, (5) industrial health records, and, in some African countries, (6) school health records and (7) sample surveys.

As a general rule, the statistics are found in publications of annual vital statistics and in statistical bulletins. These may appear monthly or annually. However, the specific conventions in each country lengthen the list to varying extents.

Apart from local sources that give data on various aspects of health, international sources also provide further data. The World Health Organisation (WHO) issues a *Yearbook on Epidemiological and Vital Statistics* for member countries, which for the most part does not duplicate the United Nations's *Demographic Yearbook*. These sources of data are valuable in building up morbidity statistics for the country or regions.

Uses of Morbidity Statistics

Morbidity statistics are collected for a number of different purposes. First, they can be used for the control of epidemics. Second, they can be used for local administrative action where infectious cases have to be identified and isolated. Third, the statistics can be used for medical intelligence systems or for research and diagnostic study. Fourth, statistics can be used for international comparisons.

The WHO collects comparable international returns on certain diseases for this purpose. However, for the most part, the scope and validity of morbidity information vary considerably from country to country. Mortality rates also serve as indicators of health conditions over much of the world.

TYPES OF HEALTH DATA

The uses mentioned above are to some extent related to the types of data collected from the system. We therefore now discuss data under the five main headings mentioned above.

Population Data

Population data are a basic requirement for most planning purposes, and data for most African countries are likely to come from censuses and/or a number of surveys. Data from another important source – civil and vital registration systems – are either non-existent or, where the data exist, defective or incomplete or both. Demographic data are indispensable to health-planning. Because planning services are for people, we need to know the total number of persons in the country and in each district or area of the country, distribution of the population, migratory movements, urbanisation, and so on. Knowledge of the basic statistics on these topics is important for the location of health facilities.

The statistics on the age–sex composition or distribution are equally important, because the health problems and needs of the various age groups are different; for example, in African countries about 45 per cent of the total population is under 15 years of age (Population Reference Bureau, 1975); and as we all know, young children are particularly vulnerable to certain diseases. The risk of death or the probability of a new-born baby's dying before the age of 5 years is extremely high; mothers and pregnant women are also exposed to additional risks. These risks require the provision of maternal and child health services for this group of persons. Young adults, on the other hand, usually require relatively little medical attention, and mortality is at its lowest level for this group in almost all populations.

Statistics on the number of births in a given period are of considerable interest in connection with the need for antenatal and postnatal services. In most developing countries the percentage of children born inside hospitals or clinics is extremely small; and if data are available on hospital deliveries by area, the assessment of the unmet needs becomes easier. Furthermore, if data are available at periodic time intervals, it becomes easier to assess the extension of maternity services to the population over time. Data on abortions, whether spontaneous or induced, although difficult to obtain, are used in making decisions on improving family-planning services and maternity services.

Statistics on mortality are usually presented by age and sex. This is important, because mortality varies with age; for example, two countries, one a developing nation and the other already developed, may have the same crude death rate but very different health conditions, age structures and age-specific death rates. High infant and child mortality rates are found in most African countries, with levels well over 100–300 infant deaths per 1,000 live births (see, for example, Population Reference Bureau, 1975). However, the early childhood mortality at ages 1–4 is less high than infant mortality rate; but if we compare these rates with the rates in developed countries, we find that they are 30–40 times the rate of developed countries (Cantrelle, 1975). The reduction of these high rates of infant and child mortality constitutes one of the challenges to health-planners in developing countries. Within the same country, death rates may vary between different social groups and between different communities.

In general, the following types of population data are needed or required:

(1) Population distributions (urban and rural, major cities and localities), for the calculation of rates of population growth and with vital statistics for measuring intercensal net loss or gain through internal migration. This information forms an essential part of any subsystem of sociodemographic statistics.
(2) Age – sex distributions, with vital statistics for measuring mortality, morbidity rates and life table functions.
(3) Major groups of birth and sex of the population, for studies on the health status of migrants and their special needs for services and facilities.
(4) Literacy or level of education by sex and age groups, for assessing the educational status of the population as an aid to action in the public health field.
(5) Economically active population by occupation, sex and age groups, for studies of occupational health hazards in relation to the health status of various occupation groups and for studies of occupation as a factor in mental health.

Health Problems Data

Health problems data relate mainly to data on morbidity or illness.[1] However, development of such statistics has been slow, but this slow development has been due to the facts that sickness is not statutorily reportable and that the early activities of public health administration were confined to the control of infectious diseases with the help of data on reportable diseases and mortality statistics. The collection of mortality statistics is easier, since death is a definite event, occurring at a definite point of time, whereas morbidity is not distinguishable from acceptable good health.

In developing countries, children and youths constitute a sizable proportion of the population, and many of the communicable diseases, such as poliomyelitis, diphtheria and smallpox, have special relevance to the younger groups of population. For the control of diseases and evaluation of public health services, collection of morbidity statistics in respect of children and youths, in particular, is an essential prerequisite. Two of the potential survey methods are health interview survey and diagnostic survey.

The former envisages the following objectives: first, the provision of data on the incidence and prevalence of illness; second, the provision of data on the utilisation of medical care facilities; third, the provision of information on the immunisation status of the population covered; and finally, the provision of information on related economic and social factors.

Accordingly, the statistics obtained in morbidity surveys are useful in a variety of ways: first, for the evaluation of the health status of the population; second, for the investigation of factors influencing health; third, for the evaluation of the effectiveness of health measures and the assessment of future needs; and finally, for the study of utilisation of health services.

Health Services Resources

Health-services input data deal primarily with the resources available for health services, and the extent to which they cover the population and the number of persons who have reasonable access to these services. The statistics required usually cover the number of hospitals and their locations, the number of hospital beds, health centres and clinics, the health manpower (doctors, nurses, etc.), training rates of health manpower and the total employment in the health sector. These input data are very scanty in most developing countries, and the statistics should normally be available; but if these are not available, a survey should be conducted to collect the information and provision made to update it regularly.

Like morbidity statistics, the statistics on health services and facilities serve a number of purposes. First, the data can be used for assessment of the present and future requirements for health services and facilities in the light of the needs of specific age – sex categories of the population. Second, the data can be used for the assessment of health problems and needs of various social and economic groups. Third, they can be used for the evaluation of the effectiveness of services and facilities already provided. Finally, they can be used for the evaluation of effects of internal migration and other aspects of social and economic change upon the health status of the population, needs for service and differential use of facilities.

The health facilities in a country in most developing countries are normally plotted on a map that shows the main roads and transport network in each district or region of the country. The purpose of this exercise is to relate the number of health care facilities to the population of the district, to facilitate comparison.

A number of crude measures can be derived from the statistics. Hospital beds are a measure of the capacity to treat inpatients, whereas the number of primary health-care facilities is an indicator of the population's access to some form of medical care. These crude measures or ratios often reveal, in a simple but effective way, the existence of large inequalities in the distribution of health services in the country.

The distribution of health manpower in most developing countries is uneven, indicating in most cases a high concentration of skilled manpower in the urban areas. The distribution pattern is very important in relation to health input, because the service is on a person-to-person basis and the doctor and the patient have to be in the same place.

Health Services Achievements

Health-services output data generally include hospital attendance, patient days, outpatient attendances, deliveries attended by qualified midwives, antenatal clinic attendances, immunisations, inpatient disease patterns, outpatient disease patterns and variations in the utilisation of health facilities. These data are usually collected by sending an already designed form to hospitals and clinics in the country to furnish the health statistics office with the information, which can be compiled on a regional and national basis for use by health administrators or planners.

Health Services Costs

The most common approach to health-planning and practice has been what may be termed the 'health budget' approach of estimating demands or needs for health services. This approach selects certain health targets – such as maternity care for pregnant women, prenatal and postnatal infant care – and, given the matrix of kinds of health personnel needed for each type of person service, the total amount of time in doctor-hours and so on is obtained by multiplication; this time multiplied by the expected cost per hour yields the anticipated cost of the health services. The data therefore are usually collected on total expenditure on health by health agencies, revenue collected, costs of hospital inpatient services, cost of outpatient services, cost of health training, cost of pharmaceuticals and expenditure by areas. Geographical distribution of health expenditure may be estimated from these data and other indicators needed for planning. Table 3.1 gives details of health statistics data needed for planning purposes.

AVAILABILITY OF DATA ON HEALTH IN AFRICA

Data on health resources are available in most African countries (see the WHO *Yearbook on Epidemiological and Vital Statistics*). The

Table 3.1 *Health statistical data*

Important and generally available	*Not so important or less available*
Demographic Data	
Population by areas	Gross reproduction rate
Age – sex distribution	Abortions (spontaneous and induced)
Crude birth rate	Stillbirths
Crude death rate	Neonatal deaths
Infant mortality rate	Registration of births and deaths
Early childhood mortality rate	
Urbanisation and other population movements	
Health Problem Data	
Some data on major health problems	General morbidity survey
	Deaths in hospitals
Health Services Input Data	
Hospitals (location)	More detailed information
Hospital beds	
Health centres (and clinics)	
Health manpower working	
Training rates of health manpower	
Total employment in the health sector	
Health Services Output Data	
Hospital admissions	
Patient days	
Outpatient attendances	
Deliveries attended to by qualified midwives	
Antenatal clinic attendance	
Under-5 clinic attendance	
Immunisation	
Inpatient disease patterns	
Outpatient disease pattern	
Variation in utilisation of health facilities	
Health Services Costs	
Total expenditure on health, by health agencies	
Revenue collected	
Costs of hospital inpatient services	
Cost of outpatient services	
Costs of health training	
Costs of pharmaceuticals	
Expenditure by areas	

Source: Brewis (1975).

series is published annually; and apart from information on the numbers of hospital beds, doctors, health workers, and so on, simple indicators such as population per hospital bed, population per physician and population per non-physician primary health worker are provided. These data are available for Nigeria, Ghana and Sierra Leone. Data on measuring health status – that is, the gross national product (GNP),[2] crude birth rate, crude death rate, infant mortality and the expectation of life at birth – are available on almost all countries of the African continent.

Information on health expenditure is available for Upper Volta, Uganda, Togo, Kenya, Sierra Leone, Ghana, Senegal and Liberia; these usually give data on the health budget as a percentage of the national budget, the health budget as a percentage of the GNP and government health expenditure *per capita*. Information is also available for Tanzania, Kenya and Tunisia on government health expenditure by public health, curative care, training and research, as well as expenditure on inpatient and outpatient services.

Data on hospital utilisation is available by year, total beds, discharges, patient days, average length of stay and occupancy rate (per cent). The distribution of medical doctors by urban and rural, and by capital city and the remainder of the country, is given for Kenya, Ghana, Senegal and Tunisia, as well as the analysis of government health expenditure.

METHODS OF ANALYSIS OF HEALTH DATA

In general, the method of analysing health data follows the pattern of deriving rates, ratios and other indicators, which, although simple to calculate, tell the administrator, in a very effective way, the nature of health problems in any country. Two rates are usually employed in this method of measurement and analysis of health. These are commonly referred to as incidence and prevalence rates. The incidence of a disease is the rate at which people are contracting the disease during a year. It is normally used in cases of acute diseases such as measles, lasting for only a short time. On the other hand, prevalence of a disease is the proportion of persons that have the disease at a particular time, and the term is usually used in connection with chronic illness.

There are several uses to which data on hospital admissions can be put; for example, patient-days is the number of days spent by patients in hospital; 100 patients in hospital for a week is equivalent to 700 patient-days, and so on. Simple relationships can be useful; for example,

$$\text{Number of admissions} = \text{Discharges} + \text{Deaths}$$

If we divide patient-days by admissions, we obtain average length of stay; that is,

$$\text{Average length of stay} = \frac{\text{Patient-days}}{\text{Admissions}}$$

Another example is occupancy rate, which is the proportion of beds occupied. This can be estimated as

$$\text{Occupancy rate (\%)} = \frac{\text{Patient-days}}{\text{Bed-days}} \times 100$$

where the number of bed-days is equivalent to the number of beds multiplied by the number of days in the year. Finally, a crude measure of occupancy is the turnover. Turnover is defined as

$$\text{Turnover rate} = \frac{\text{Admissions}}{\text{Beds}}$$

In developing countries where patients sometimes sleep on the floor, this implies that the capacity of the hospital has been exceeded, and the occupancy rate is then given as over 100 per cent.

The importance of these rates for planners cannot be overemphasised; for example, the average length of stay is usually regarded as a measure of the hospital's management efficiency, but this rate says nothing about the effectiveness of treatment in the hospital. High occupancy rates are used as an argument for building extra capacity for the hospital. Information on outpatient attendances gives an estimate of illness and an index on the demands being made on the health services or indicates the workload of the staff. Rate of utilisation of health facilities can be estimated from the population of a catchment area and the number of attendances. Naturally, if there are great variations in this index within a country, this suggests inequalities in the distribution of health care resources.

In most areas, fairly crude measures of output may be acceptable, but the measures are naturally subject to the usual index-number problems. These estimates are approximate, because they are based on incomplete data. A problem with the indicators is the criteria for their choice and selection. In general, they should be relevant, representative and sensitive and should either positively or negatively be correlated with other dimensions of health. Finally, they should be quantifiable and should relate to outputs of health services rather than inputs.

LIMITATIONS OF HEALTH STATISTICS

The statistics on health, like statistics from other sectors, are subject to several limitations. They vary from incomplete coverage to details in the tabulations provided by various countries. There is also lack of uniformity in data collection procedures and the publication of such information. Complaints about delay in publishing statistics are, of course, an occupational hazard, but it affects certain branches of statistics more than others. Some of these limitations can be traced to the nature of the scope and source from which the statistics are derived, as well as the definition of 'health'. For example, the notification of diseases, which began in the early days of public health administration, is mainly confined to communicable diseases and cannot be considered dependable, because its efficacy depends not only on the extent of morbidity in a community but also on several other factors, such as the extent of peripheral health services, co-operation from the medical profession, nature of the disease and willingness of the patient. As the extent of undernotification is high and often unknown, the statistics collected are essentially useful only for the operational purposes of controlling diseases by taking immediate steps for prevention and disease surveillance, and for rough assessment of trends in the incidence of the disease.

The statistics, however, suffer from various limitations, the most important being coverage and selectivity. Hospital statistics are, no doubt, an accurate source of diagnostic information, but they suffer from bias arising from selectivity in relation to factors such as type and severity of disease, provision of health facilities, age and sex of the patient, and social and economic factors. The population served is also unknown. It is not, therefore, possible to generalise the hospital experience in respect of many diseases to community level, and these statistics cannot give a true picture of prevalence or distribution of morbid conditions in a community as a whole. However, they serve the purpose of measuring the relative distribution of diseases in the areas covered, and can be treated as valuable adjuncts to mortality statistics, suggesting priorities for provision of more medical facilities and efficient medical care.

Records maintained for physical examination of school children constitute another source from which morbidity patterns can be ascertained. But in many developing countries the school medical-examination system may not cover all the schools, and the disease recorded may not be specific. This source, under the existing conditions, may not be useful, even for school children, who constitute only a segment of the entire population of children and youths. Another important source of morbidity statistics is the records of general practitioners. Examples of studies based on such records are the study conducted in England and Wales during 1955 – 6, and the ones conducted on a national basis. Such studies may not be useful in developing countries, since many of the causes of illness may go unattended by qualified doctors. The other sources of morbidity statistics, such as sickness insurance records and records of sickness and absenteeism in industry, are only of restricted utility, since they relate to specific groups of population. In fact, they are incidental to the operation of some services and are not meant for building up statistics for the operation of health services as such.

The use of health statistics for different countries is often difficult. For comparative studies,the data may require considerable correction and adjustment as well as a great deal of handling. The existence of international summaries of health statistics, while not entirely eliminating the necessity for researchers in national collections of statistics, considerably facilitates in cases where the details required are too expensive.

CONCLUDING REMARKS

We end this chapter with the recognition that what statistics are available both mould and also reflect our perception of reality. We are restricted by the availability of data, but it is worth emphasising that the five categories of data on health described in this chapter, if available even in scanty form, and if properly organised, will be useful for description, analysis and policy-making in the health sector of the country. They may also provide a basis for the range of internationally comparable statistics in the general area of social statistics. The value of this kind of comparability for analytical purposes can hardly be exaggerated, since, taken together, the countries of the world exhibit, in most matters, far more variability than can be found in the data for a single country.

Finally, the development of morbidity statistics is rather uneven, for several reasons, and standard methodology of a survey method in relation to the conditions in Africa has still to be developed. Nonetheless, the need for such surveys is compelling and brooks no delay. The simultaneous development of mortality and other demographic statistics, with due emphasis on causes of death, need to be emphasised. The improvement of normal registration systems, introduction of medical certification in the international form, and building up of estimates for vital rates are some of the possibilities in the offing. But the consciousness of the need for morbidity statistics and the necessity for building up adequate organisation, with sufficient funds made available through development plans or otherwise, should be generated in the government agencies responsible for vital and health statistics.

NOTES: CHAPTER 3

1 Morbidity is a technical term, and like crime it is difficult to define. Good health can be defined as a state of complete physical, mental and social well-being, and not merely the absence of disease or infirmity.
2 Gross national product (GNP) is defined as the market value of the total output of goods and services produced by a nation's economy.

REFERENCES AND FURTHER READING

Akinla, O. and Adadevoh, B. K. (1969), 'Abortion: a medico-social problem', *Journal of the Nigerian Medical Association*, vol.3, pp.16 – 22.

Ayeni, O. (1976), 'The importance of morbidity statistics in the evaluation of public health in Africa', *Jimlar Mutane*, vol. 1, no. 2, pp. 193 – 7.

Brewis, F. (1975), 'Health statistics', paper presented at the conference on Statistical Policy in Less Developed Countries, Institute of Development Studies, University of Sussex, Brighton (12 – 16 May), IDS No. 114.

Brotherston, J. H. F. (1973), 'Health planning and statistics: an overview from Scotland', *International Journal of Health Services*, 3:35.

Cantrelle, P. (1965), 'Mortalité: facteurs', in Délégation Générale à la Recherche Scientifique et Technique, *Demographie Comparée: Afrique Noire, Madagascar et Commores*, pt 5.

Cantrelle, P. (1975), 'Mortality: levels, patterns and trends', in J. C. Caldwell *et al.* (eds), *Population Growth and Socio-Economic Change in West Africa* (New York: Population Council Books).

Clarin, R. (1968), 'The assessment of infant and child mortality from the data available in Africa', in J. C. Caldwell and C. Okonjo (eds), *The Population of Tropical Africa* (London: Longman).

Conran, A. and Conran, O. F. (1956), 'Medical survey of Tonkolili and adjacent valleys', *Journal of Tropical Medicine and Hygiene*, 59, pp. 285 – 94.

Cummings, E. C. (1971), 'The development of health services in Sierra Leone', in *The Demographic Transition in Tropical Africa*, proceedings of an Expert Group meeting, Paris, 17 – 19 November 1970 (Paris: Development Centre, Organisation for Economic Co-operation and Development), pp.253 – 7.

Cunningham, N. (1968), 'An evaluation of an auxilliary-based child health service in rural Nigeria', paper presented to the annual congress of the Nigerian Society of Health, University of Ibadan, Ibadan.

Dawson, J. L. M. (1964), 'Urbanization and mental health in a West African community', in A. Kiev (ed), *Magic, Faith and Healing: Studies in Primitive Psychiatry Today* (New York: Free Press, pp. 305 – 42.

Dixon-Fyle, S. R. (1973), 'Climate, health and the efficiency of labour in tropical Africa', paper presented at the conference of the Planned Parenthood Association of Sierra Leone, Freetown.

Gale, T. (1976), 'The struggle against disease in Sierra Leone: early sanitary reforms in Freetown', *Africana Research Bulletin*, vol. 6, no. 2, pp. 29 – 44 (Freetown: Fourah Bay College).

Gamble, D. P. (1961), 'Infant mortality rates in a Sierra Leone urban community (Lunsar)', *Journal of Tropical Medicine and Hygiene*, 64, pp. 192 – 9.

Ghana (1970 – 8), Results (still coming out) of a series of surveys on morbidity in connection with the Danfa Project, carried out by the Department of Community Health.

Ghana (annual), Reports on medical statistics, Biostatics Section, Ministry of Health, Accra.

John, G. (1975), 'A comparative analysis of causes of death for selected countries', *Regional Institute for Population Studies Newsletter*, no. 5, pp. 6 – 20.

Johnson, O. G. (1953), *Records and Reports of Local Health Departments*, Public Health Monograph No. 14 (Washington, DC: US Department of Health, Education and Welfare).

Kamara, M. P. A. (1979), 'The early history and the present registration of births and deaths in Sierra Leone', Office of the Registrar of Births and Deaths, Freetown (mimeo).

Kandeh, H. B. S. (1978), *Infant and Child Mortality in the Bo District of Sierra Leone (First Report)* (Njala: Ford – Rockefeller Population Policies Research Programme, Njala University College, University of Sierra Leone).

Kpedekpo, G. M. K. (1970), *Studies on Vital Registration Data from the Compulsory Registration Areas of Ghana, 1962-67*, Vols 1 and 2, (Legon, Accra: Institute of Statistical, Social and Economic Research, University of Ghana).

Lagos, (1972), 'Vital statistics: Lagos division or city, 1968 – 71', State Government, Ministry of Health and Social Welfare, Medical Statistics Division (mimeo).

Makannah, T. J. (1977), 'Births, deaths, and population and distribution in the western area: a historical review', *Sierra Leone Quarterly Health Education Journal*, no. 1 (4).

McCuster, J. (1976), 'Morbidity in Tanzania', *Jimlar Mutane*, vol. 7, no. 2, pp. 198 – 207.

McGregor, I. A., Billewicz, W. Z. and Thompson, A. M. (1961), 'Growth and mortality in children in an African village', *British Medical Journal*, pp. 1661 – 6.

Mills, A. R. (1962), 'The effect of urbanization on health in a mining area of Sierra Leone', Department of Social Medicine, University of Edinburgh (mimeo).

Mills, A. R. (1967), 'A comparison of urban and rural populations in the Lunsar area: 1 Lunsar', *Sierra Leone Studies*, new series, no. 20, pp. 173 – 90.

Mills, A. R. (1967), 'A comparison of urban and rural populations in the Lunsar area: 2 village populations', *Sierra Leone Studies*, new series, no. 21, pp. 12 – 51.

Mills A. R. (1967), 'The effect of urbanization on health in a mining area of Sierra Leone', *Transactions of the Royal Society of Tropical Medicine and Hygiene*, 61, pp. 114 – 30.

Mills, A. R. (undated), 'A method of assessing health in an African community, and its application in assessing the effects on health of urbanization', unpublished paper.

Nair, G. V. (1973), *Medical Statistics in Sierra Leone, 1969 – 1971* (Freetown: Ministry of Health).

Neumann, A. K. *et al.* (1972), 'The Danfa – Ghana Comprehensive Rural Health and Family Planning Project: preliminary report', *Ghana Medical Journal*, 11:18.

Nickson, R. A. (1974), *Allocation of Health Resources in Sierra Leone*, (Njala: Njala University College).

Nigeria (1953), *Annual Report, 1952 – 3* (Department of Medical Services of Northern Nigeria).

Nigeria (1954), *Annual Report, 1953 – 4* (Department of Medical Services of Northern Nigeria).

Nigeria (1963), *Annual Statistical Bulletin* (Ibadan: Ministry of Health, Western State).

Nigeria (1968), *Annual Statistical Bulletin*, (Ibadan: Ministry of Health).

Nigeria (periodic), *Statistical Abstracts of Nigeria* (Lagos: Federal Statistics Office).

Ogunmekan, G. O. (1973), 'Analysis of medical admissions to Adeoyo State Hospital, Ibadan, 1969', *Nigerian Medical Journal*, 3, pp. 5 – 12.

Page, J. H. and Coale, A. J. (1969), 'Estimates of fertility and child mortality in Africa south of the Sahara', paper presented to the seminar on Population Growth and Economic Development, Nairobi University, Nairobi.

Population Reference Bureau (1975), *World Population Data Sheet* (Washington, DC, annual).

Sai, F. T. *et al.* (1972), 'The Danfa – Ghana Comprehensive Rural Health and Family Planning Project: a community approach', *Ghana Medical Journal*, 11:9.

Sierra Leone (1926), *Annual Medical and Sanitary Report for the Year Ending 31st December, 1925* (Freetown: Medical Department, Ministry of Health).

Sierra Leone (1962), *Development Plan of Health Care Services of Sierra Leone* (Freetown: Ministry of Health/Government Printer).

Sierra Leone (1964), *Report of the Medical and Health Services* (Freetown: Medical and Health Services/Government Printer).

Sierra Leone (1964), *Report of the Medical and Health Services, 1959* (Freetown: Medical and Health Services/Government Printer).

Sierra Leone (1970), *Proposal for the Expansion of Civil Registration in Sierra Leone* (Freetown: Central Statistics Office).

Sierra Leone (1973), *Vital Statistics and Causes of Death: Western Area, 1969 – 71* (Freetown: Medical Statistics Unit, Ministry of Health).

Sierra Leone (1977), *Vital Statistics and Causes of Death Report: Western Area, Sierra Leone, 1971/72 – 1975/76* (Freetown: Ministry of Health).

Tenney, J. B. (1973), 'Information for developing national statistics on ambulatory medical care', *Medical Care*, 11: (supplement) 87.

Tekse, K. (1970), *Some Estimates of Vital Rates for Sierra Leone*, World Health Organisation Epidemiology Project (Project Report) (Freetown).

United Nations (annual), *Demographic Yearbook* (New York: UN Statistical Office, Department of Economic and Social Affairs).

United States (1978), *World Population, 1977: Recent Demographic Estimates for Countries and Regions of the World* (Washington, DC: Bureau of Census, Department of Commerce).

United States National Academy of Sciences (undated), *Public Health Problems in Fourteen French-Speaking Countries in Africa and Madagascar*. Vol. 7: *A Survey of Resources and Needs* (Washington, DC: Division of Medical Sciences, National Research Council).

World Health Organisation (WHO) (1967), *Third Report on the World Health Situation, 1961 – 1964*, Official Records No. 155 (Geneva).

WHO (1969), *International Classification of Diseases*, 2 vols (Geneva).

WHO (1971), *Statistical Indicators for the Planning and Evaluation of Public Health Programs: Fourteenth Report of the WHO Committee on Health Statistics* (Geneva).

WHO (annual), *Yearbook on Epidemiological and Vital Statistics* (Geneva, published for member countries of WHO).

Wilde, H. (1968), 'Child malnutrition in Sierra Leone', report prepared for the US Department of State, Washington, DC.

Wilkinson, J. L. (1962), 'Survey of infant mortality in the eastern province', *The Bulletin: The Journal of the Sierra Leone Geographical Association*, 6, pp. 2 – 3.

Wilkinson, J. L. (1964), 'Child malnutrition in Sierra Leone', *Scottish Medical Journal*, vol. 9, pp. 64 – 7.

Wilkinson, J. L. (1964), 'Ecology and prevention of child malnutrition in a rural community', *West African Medical Journal*, vol. 13, no. 1, pp. 9 – 18.

Wilkinson, J. L. (1965), 'Measurement and control of mortality among children in Sierra Leone and other West African countries', *Journal of Tropical Medicine and Hygiene*, 68, pp. 167 – 73.

Wilkinson, J. L. (1969), 'Children in hospitals in Sierra Leone: a survey of 10,000 admissions', *Transactions of the Royal Society of Tropical Medicine and Hygiene*, vol. 63, no. 2, pp. 263 – 9.

Williams, B. E. O. (1971), 'Maternal mortality in the Princess Christian Maternity Hospital, Freetown, Sierra Leone', unpublished data, Ministry of Health, Freetown.

Williams, B. E. O. (1973), 'Study of abortion in Freetown', unpublished data, Ministry of Health, Freetown.

Williams, B. E. O. (1975), *Annual Report, 1974 – 5* (Freetown: Maternal and Child Health Division, Ministry of Health).

Williams, B. E. O. (1976), *Annual Report, 1975 – 6* (Freetown: Maternal and Child Health Division, Ministry of Health).

Williams, B. E. O. (1977), 'Population and health', paper presented at the First National Population Seminar, sponsored by United Nations Fund for Population Activities/UNESCO, Fourah Bay College, University of Sierra Leone, Freetown (14 – 22 December).

Williams, B. E. O. *et al.* (1972), 'Report of the Sierra Leone delegation on the health situation in Sierra Leone', paper presented to a regional conference on Labour and Family Health Seminar, Bathurst (mimeo).

Williams, G. A. (1969), *Annual Report of the Medical Officer of Health Lagos, for the year 1968* (Lagos).

CHAPTER 4

Educational Statistics

The importance of people not only as the ultimate beneficiaries but also as the mainspring of the whole process of economic growth obviously implies that education and training must have a prominent place in every development plan. Recently, many governments in Africa have been emphasising the need to eliminate illiteracy, to provide universal and free primary education, and to provide enough manpower with technical and academic skills to meet the requirements of development plans. Governments also realise that, by easing the shortages of manpower, education can reduce the need for high salaries for those with scarce professional skills, and thereby reduce inequalities in income.

However, the administration of an educational system depends to a great measure upon the production of statistics, both in coverage and in sufficient quantity and quality for use in evaluating the system and in planning for its future development. This chapter therefore deals with the nature and scope of educational statistics, sources, their sources and limitations. The statistics treated are those obtained from the schools and also literacy statistics obtained from censuses and surveys.

SOURCES OF EDUCATIONAL STATISTICS

It is convenient to begin the description of the sources of educational statistics by giving a detailed description of one country, Ghana, whose coverage of educational statistics is comprehensive, and then to generalise on sources from this description. As early as 1893 the annual report of the then Educational Department contained statistical information about the educational system. In 1954 a separate booklet, *Education Statistics*, was issued, and extracts from this new publication were inserted in the Ministry of Education annual report. After 1959 the *Statistics* was discontinued, and since then the only detailed published educational statistics have been those issued once in every five years by the Central Bureau of Statistics (CBS) in collaboration with the Ministry of Education. The CBS annual *Statistical Year Book* contains educational statistics of a less detailed nature. The ministry's own annual statistical tables are bound up in loose form and distributed to members of the ministry who have immediate use for them. All others are supplied with information on request.

Prior to 1968, the monthly pay vouchers were used as the source of information on primary and middle school pupils and teaching staff and their qualifications. The information in the salary vouchers was extracted and summarised in the form of tables in the district education offices. They were sent to the regional education offices, where they were compiled into files and sent to the Statistical Unit. These files contained a list of all educational institutions in the region.

Statistical information on secondary schools, commercial schools, technical institutes, teacher-training colleges and all types of private educational institutions was obtained from completed (mailed) questionnaires submitted to the Statistical Unit by heads of these educational institutions. In the 1968–9 school year, it was decided that the postal census method (mailing of designed questionnaires) should be used to cover all educational institutions. Questionnaires to the primary and middle schools were routed through the regional education offices, and each primary and middle school was asked to attach to the completed questionnaire a copy of the salary voucher completed on the census date. These were collected through the regional education offices, which returned their completed forms direct to the Statistical Unit, serving the district and regional officers with copies. Data on the universities were obtained through the Higher Education Division of the ministry (now the National Council for Higher Education).

The census date is the last day for classes in November of any year (28 November in 1968). This date is chosen as representing the earliest practical moment in the new school year when schools have settled down with a stable body of staff and pupils. The format for the educational statistics publications issued on an annual basis by the Planning Unit of the Ministry of Education, started from 1968–9, consists of the following essential features.

Part I deals with educational statistics of primary, middle, secondary and commercial schools, teacher-training colleges, technical institutions and universities. It generally gives numbers, classes and type as well as grade. Part II gives time series data on the growth in number of institutions, growth enrolment, growth of teaching staff, examination performance and finance of education. Finally, a section is given depicting the structure of the educational system in Ghana, the structure of the commercial and technical education system and the organisation of the Ministry of Education.

The tables are by far the most comprehensive on the educational system of an African country. A major omission from these tables, however, is data on the ages of pupils by grade, but these data are collected by the Central Bureau of Statistics in its quinquennial census and published in its *Statistical Bulletin*.

In general, therefore, the main sources of education statistics are the records of:

(1) schools and universities
(2) local and regional education offices;
(3) departments within the Ministry of Education;
(4) other government departments (Central Statistical Office); and
(5) in addition, special inquiries that can be undertaken.

The collection of these statistics is usually done by employing three methods. These are (1) personal inquiry, (2) postal census, and (3) sample inquiries. As a general rule, the statistics are found in the statistical bulletins or abstracts and Ministries of Education of African countries.

COVERAGE OF EDUCATIONAL STATISTICS

In order to provide the basic information required for evaluating and planning in a sufficiently comprehensive and accurate fashion, the following items are required:

(1) Statistics of schools should show, by type of management and for each level, (a) number, (b) size and location, and (c) equipment.
(2) Statistics of teachers should show their numbers by sex, general qualifications, whether full or part time (in each type of school and at each level), the subjects taught, details of resignation and death.
(3) Statistics of pupils (students) should show their numbers by sex and age (in each grade, in each type of school and at each level), subjects studied, transfers and deaths, and successful leavers.
(4) Financial statistics should be available to make it possible to estimate the cost of each pupil, teacher and building unit at each level and the total annual recurrent and capital expenditure at each level and for administration.

The sources of inaccuracy are many and can include: lack of completeness of the records, delays in notification and the supply of inaccurate information.

LITERACY DATA ON EDUCATION FROM CENSUSES AND SURVEYS

Advances in literacy and in economic development are interdependent and reinforce one another. Accordingly, efforts to organise local planning and to foster the growth of co-operative movements are critically dependent on the extent to which functional literacy is diffused through all sections of rural and urban life. Recognising these relationships, a number of African countries are making strenuous efforts to obtain information on the literacy level of the population from censuses and sample surveys, but conceptual and definition problems have negated most of these efforts.

In such surveys, the definition of literacy given by the United Nations Educational, Scientific and Cultural Organisation (UNESCO) is adopted. It defines a literate person as 'one who can with understanding both read and write a short simple statement on his everyday life in any language' (UNESCO, 1961, p. 40). Accordingly, all persons can be divided into two groups: persons who can read and write; and all persons who cannot read nor write, those who can only write, and those who can read and write figures and their names only. This definition of literacy is vague and thus indicates persons with only a modicum of reading and writing ability as literate. In the absence of a specific literacy test applied at the time of the survey, literacy results obtained from such surveys can be misleading.

For example, in Tanzania, of the educational questions in the 1967 population census, the one on literacy gave results remarkably similar to that on formal education. In the population at large, schooling, however short, meant that the person was literate, and absence of formal education that he was illiterate. Attempts to test the level of literacy a bit more accurately in a subsequent household budget survey also failed, and it can therefore truly be stated that we have no useful estimate of literacy in Tanzania (Egero, 1973). As a result of these difficulties there are few countries in Africa with any reliable data on functional literacy.

USES OF EDUCATIONAL STATISTICS

There are two groups of people who need educational statistics as tools for decision-making: the administrators and the planners.

Statistics and Administration

In most developing countries government expenditure on education accounts for 10–25 per cent of the annual budget; education therefore is a large scale operation, and detailed financial statistics are required for accounting purposes. Second, the Ministry of Education is responsible for registering in its schools as much as 15 per cent of the population, and for employing teachers and other personnel; as a result, the educational system generates a considerable supply of statistics on pupils, teachers and other personnel, buildings and finance.

Statistics and the Evaluation of the Educational System

The statistics generated and collected on the educational system can be used to provide an overall statistical picture of the extent and working of the educational system. Several types of indicators are required, and the statistics provide quantitative answers to questions such as:

(1) How large is the system?
(2) What kinds of education is it providing?
(3) How rapidly is it growing?
(4) Which areas or regions are lagging behind in educational development?
(5) What use is being made of teachers and the schools?

Adequate answers are usually required to these questions in order to justify the budgeting expenditure, to make estimates for future expenditure and to plan future development.

Educational Statistics and Planning

The statistical evaluation of the educational system is necessary for educational planning. The present educational facilities represent the stock immediately available for future development; together with flow statistics, they provide a good basis for indicating how the educational system is working. A discussion of planning methods is given in a different section.

We may therefore summarise the uses of educational statistics as follows:

(1) The statistics on education provide an overall statistical picture of the extent and working of the educational system.
(2) The statistics also help to evaluate the performance of the educational system.
(3) The statistics are necessary for future planning through projections of the future school population and supply of teachers, and so on.

(4) The statistics are useful for planning the manpower requirements for the country.
(5) The statistics can be used for international comparisons.

Statistics collected on education on a worldwide basis by international agencies aim at standardising the amount and form of educational statistics published by member governments. An incidental, but highly important, effect of these activities is the effort involved in checking national definitions and coverage. This may reveal where gaps and ambiguities exist, and these may lead to clearer thinking about national statistics.

There are, however, several comparability problems at the international level. The first is that international comparisons in educational statistics are far more difficult to achieve than in many other fields, the reason being that educational systems possess strong national peculiarities arising out of the history and policies of individual countries. Second, comparability problems are further aggravated by the special conditions obtaining in most developing countries; for example, the uneven development in education, the payment of fees, the lack of reliable information on births and deaths, the wide age range found in a single grade, the high rate of absenteeism, the high rate of repeaters, and the high rate of drop-outs.

Nevertheless, simple international comparisons are attempted on the basis of the numbers enrolled at different levels by subject at the third level, enrolment ratios, educational finance and teacher supply. The main problem in comparisons of enrolment lies in the definition of what constitutes the first, second and third levels of the educational ladder. In the comparison of enrolment ratios use is made of the age-specific enrolment ratios. A simpler summary measure is also required to make such statements as that x per cent of the population of one country is receiving education compared with y per cent in another country. A simple gross enrolment ratio for all levels is unsatisfactory, and attempts to use other indicators and a standard population to which to relate enrolments are not satisfactory either.

Comparisons of cost of education are of a very limited nature, because satisfactory financial statistics are not available. The comparisons are limited to the comparison of public expenditure on formal education, because national expenditure on formal education lends itself to detailed comparison by level and purpose. Finally, the number of teachers gives a better overall indicator of a nation's educational efforts than either the number of pupils or the amount of expenditure on education. The effort is measured not in monetary terms but by the numbers of highly qualified manpower devoted to the purpose. It is also important in comparisons of this nature to measure the quality of the teachers, but national practices regarding teacher qualifications are so diverse that this presents a problem of special difficulty.

PLANNING METHODS AND THE USE OF EDUCATIONAL STATISTICS

The literature on the links between education and manpower is often based on the idea that there is some simple connection between a job and the training necessary to perform it. In some cases this is obviously true, but in others this is not true at all; nevertheless, this has a considerable bearing on manpower-planning and methods employed in several countries.

In general, mathematical techniques for planning education and manpower have a special appeal to planners because of the quantitative characteristics of educational systems. These characteristics are the student – teacher ratios and student – classroom ratios, known as technical coefficients. These relate to inputs and outputs in much the same fashion in education production functions as in economic production functions. The models generally assume fixed technical coefficients and project needs for teachers and classrooms at various levels of education, given enrolment targets.

Two methods are discussed. In the first method, the manpower-planning approach, manpower needs are converted into educational needs by assuming some pattern of future relationships between occupations or skill levels and educational requirements. The procedure assumes ability to predict, more or less accurately, future changes in the patterns of occupational and skill demands or changes in educational requirements corresponding to those demands. This model has one serious limitation. In economics, it is generally recognised that input – output coefficients are likely to change over time, because they depend on changing conditions of supply and demand. An exactly similar situation holds in this method of approach, but the method uses fixed input – output coefficients, which is a serious limitation. In these models, population is assumed to play a role only on the supply side, and costs of producing the educated manpower are ignored in these approaches.

The second approach is the rate-of-return analysis. In this technique, differences in expected future earning streams of different educational groups are calculated with controls for various factors, such as age, sex, family background and urban – rural location. Given these differences and with adjustments for mortality and unemployment, rates of return to different levels of education are readily calculated from age – education earnings statistics obtained from household surveys or other sources. Similar methods generally assume a fixed future pattern of wages corresponding to different educational levels.

The rate-of-return method has a number of limitations. The first is that for government decision-making the appropriate comparisons concern social rates of return rather than private rates of returns; but given the imperfections in factor markets, it cannot be assumed that private sector earnings reflect social values. Second, cross-sectional observations of age – education earnings differences that are usually used to impute time series earnings are seldom statistically controlled and perhaps impossible to control (e.g. ability and motivation).

CONCLUDING REMARKS

The kind of educational and manpower-planning adopted in African countries depends on the type and quality of data available, the uses to which the results will be put and the nature of the educational system. Obviously, more use can be made of available data if uniform concepts, definitions and classifications are used in collecting data on demographic statistics, educational statistics and labour statistics. Their use will be greatly enhanced if the data on these three topics can be linked to existing bodies of data, and perhaps also to existing methods of data collection.

However, there exist serious gaps in the statistics of education collected from the various sources in African countries. Such statistics as are available are significantly more complete for formal than for informal education and for maintained than for unassisted institutions. Even the form content and periodicity of the data differ between them. Published statistics tend to concentrate too narrowly on enrolments, projected enrolments and pupil – teacher ratios, to the neglect of other vital information. For example, the data available on the costs of education are inadequate, and no solid studies seem to exist on the relative costs of education between primary, secondary and higher education, or between rural and urban institutions, between comprehensive and general secondary education, between vocational and academic education, or between vocational on-the-job training and the formal vocational education system.

There are virtually no published data in the whole field of training on the effort of private firms and voluntary agencies. A census of all training facilities (both public and private) could be undertaken, and its maintenance is highly desirable.

Rates of repetition seem to have been published for some countries but are still very few. Even these are highly suspect and tend to understate the position. It follows that the projected enrolment rates in the development plans, which seem to take no account of repeaters and drop-outs, are misleading. Data on the educational and occupational characteristics of the population, including performance in different subjects, should also become more detailed and comprehensive in order to provide a better basis for manpower forecasts. The same applies to earnings by occupation and education.

REFERENCES AND FURTHER READING

Edding, F. (1966), *Methods of Analysing Educational Outlay* (Paris: United Nations Educational, Scientific and Cultural Organisation).

Egero, B. (1973), 'Non-sampling errors', in B. Egero and R. A. Henin (eds), *The Population of Tanzania: An Analysis of the 1969 Population Census*, Census Vol. 6 (Dar es Salaam: Bureau of Statistics).

Jolly, R. (1969), *Planning Education for African Development* (Nairobi: East African Publishing House).

Kendall, W. L. (1968), 'Statistics of education in developing countries', report prepared for the United Nations Educational, Scientific and Cultural Organisation, Paris.

Sierra Leone (1971), *Statistical Bulletin*, vol. 18 (Freetown: Central Statistics Office).

United Nations Educational, Scientific and Cultural Organisation (UNESCO) (1958), *Recommendations Concerning the International Standardization of Educational Statistics* (Paris).

UNESCO (1961), *Manual of Educational Statistics* (Paris).

UNESCO (1965), *Elements of Educational Planning*, Educational Studies and Documents No. 45 (Paris).

UNESCO (1965), *Educational Planning Practice in the Developing Countries* (Paris).

UNESCO (annual), *Statistical Year Book* (Paris).

Wallberg, K. (1965), 'Statistics of the educational system and of the educational level of the population: needs and possibilities', paper presented at the conference of European Statisticians, working group on Statistics of Education, Geneva.

Young, W. C. E. (1966), *Statistics of Education, 1964–1965*, Research Bulletin No. 2 (Freetown: Department of Education, Fourah Bay College, University of Sierra Leone).

Young, W. C. E. (1967), 'Highlights of the 1964–65 statistics of education', *Journal of Education*, vol. 2, no. 1, pp. 15–18; vol. 2, no. 2, pp. 14–19 (Freetown).

MAJOR SOURCES OF EDUCATIONAL STATISTICS

ALGERIA

Ministère de l'Education Nationale, *Etats scolaires et universitaires*, nos 2 and 3 (1969).

Ministère de l'Education Nationale, *Tableaux extraits des informations statistiques: planification et orientation scolaires* (1969–70).

ANGOLA
Direcçao dos Servicos de Estatistica, *Estatística da Educaçao* (1966 – 7, 1967 – 8, 1969 – 70; annual).

BOTSWANA
Ministry of Education, *Education Statistics* (1968, 1969, 1970, 1971; annual).
Education Department, *Annual Report* (1967).

CAMEROUN
Ministère de l'Education, de la Jeunesse et de la Culture, *Annuaire statistique* (1968 – 9; annual).

CENTRAL AFRICAN REPUBLIC
Ministère de l'Education, *Statistiques scolaires* (1966 – 7; annual).

CHAD
Ministère de l'Education, *Statistiques scolaires* (1967 – 8, 1968 – 9; annual).

EGYPT
Ministry of Education, Documentation and Research Centre for Education, *Educational Trends* (1968).

GABON
Ministère de l'Education Nationale, *Statistique de l'enseignement au Gabon* (1969, 1970, 1971; annual).

GHANA
Ministry of Education, *Education Statistics* (1968-9, 1969-70; annual).
Ministry of Education, *Education Report* (1960 – 2, 1963 – 7).

KENYA
Board of Adult Education, *Annual Report* (Nairobi, 1969 – 70).
Ministry of Education, *Annual Report* (1967, 1968, 1969, 1970, 1971).
Board of Adult Education, *Journal of Adult Education*, vol. 1 (1972), vol. 2 (1973).

LESOTHO
Ministry of Education and Culture, *Annual Report* (1967, 1968).

LIBERIA
Department of Education, *Annual Report* (1968 – 9, 1969 – 70).

MALI
Service de la Statistique Générale, *Statistiques scolaires* (1968 – 9).

MAURITIUS
Department of Education, *Annual Report* (1967, 1968, 1970).

NIGER
Ministère de l'Education Nationale, *Situation de l'enseignement au Niger au 1er janvier, 1967 et 1968.*

NIGERIA
Ministry of Education, *Statistics of Education* (1967, 1968; annual).
Ministry of Economic Planning, Reconstruction and Statistics, *Annual Digest of Education Statistics: Western Nigeria* (1968, 1969).

PEOPLE'S REPUBLIC OF CONGO
Bureau de l'Enseignement Catholique, Service Statistique, *Annuaire statistique de l'enseignement catholique* (1966 – 7).

SENEGAL
Ministère de l'Education Nationale, *Statistiques scolaires* (1968 – 9).

SIERRA LEONE
Ministry of Education, *Newsletter Report Summary* (1967, 1968, 1970).
Ministry of Education, *Report* (1962, 1963, 1964, 1965, 1966, 1967, 1968, 1969, 1970, 1971; annual).

SWAZILAND
Ministry of Education, *Annual Report Summary* (1967, 1968, 1970).
Ministry of Education, *Educational Statistics* (1968, 1969, 1970, 1971, 1972).

TOGO
Ministère de l'Education Nationale, *Statistiques scolaires du Togo* (1966 – 7).

CHAPTER 5

Housing Statistics

Housing is a basic human need for the protection of persons and of property from sun, rain and wind. Housing also reflects the culture of a country and constitutes historical evidence of civilisation. However, the nature and scope of housing statistics differ from country to country. This depends on the definition of what constitutes 'housing'.

In most African countries, where a vast majority of persons invariably live in slum houses, all forms of shelter are recognised by governments as part of the housing stock. To eliminate this type of 'house' from the definition of a house means that we shall be concerned only with a few houses located in the best areas of African cities.

In both the developed and the developing countries, housing statistics describe the housing situation at a well-defined point of time (benchmark statistics) and reflect changes in the housing situation during a specified period of time (current statistics). They reflect the quantitative aspects of the housing inventory, its occupancy status and its adequacy in terms of facilities available to the occupants.

This chapter describes sources of housing statistics, uses and limitations, and problems of interpretations of the statistics.

SOURCES OF HOUSING STATISTICS

The principal source for basic housing data is housing censuses and surveys. These censuses and surveys examine housing and household conditions and facilities. Such surveys focus on: the distribution of the population by type and adequacy of housing; new family formation as a factor in the demand for housing; the estimation of housing requirements, current and future needs as indicated by population change; studies of city-planning (the number of households and persons affected in old and planned areas; and studies of housing costs per family, rents, and so on.

However, since these censuses are taken after ten years, the information on housing is updated by conducting sample surveys. These surveys have been used as an intercensal source of information and also provide more information on certain aspects of housing not covered in the complete housing census. Other sources include national income accounts, housing institutions reports and special surveys.

Data on new houses can also be obtained from construction statistics, which can be derived from permits issued for the construction of new dwellings, from records of dwellings initiated or completed and also from the certificates of occupancy. The statistics on dwellings demolished can be obtained from records maintained from property tax collection. However, in most African countries, such records are either not kept or are highly distorted.

National accounting statistics[1] can be used to estimate the contribution of houses to the national product. The method does not, however, provide information on the total number of houses. Nevertheless, the number of houses can be obtained indirectly. One method that can be used to obtain the number of houses is to go through the details of the national income computations and to find out the different categories of houses and the rents imputed to each category.

If, however, the national accounts provide data on the components of the gross and net capital formation at the current market prices, one can single out the capital formation in houses. Obviously, this will give information on the new houses only and will not give the stock of houses. Nevertheless, if a country prepares a balance sheet of the assets and liabilities of the nation, one can easily single out this number and the estimated present value of the houses.

Unfortunately, we cannot depend on this source in African countries, as the balance sheets represent only the financial assets and the financial liabilities of the banking sector. Furthermore, the detailed balance sheet for the nation as a whole (which covers both the financial and the non-financial transactions of all the sectors) is not available for African countries. The best that can be done is to take data from censuses of houses for the available years and to use capital formation in housing to build up a complete series of housing statistics.

So far we have described the general sources of data on housing. No attempt is made to give a detailed description of all sources of housing statistics in African countries, since this would not appear to be of vital interest and since the larger number of states (over forty-five) would pose something of a problem. Instead, we shall describe in detail sources[2] for one African country: Nigeria.

The main source of housing statistics now in Nigeria is the Federal Housing Authority. Earlier, the Department of Statistics initiated a series of urban consumers' surveys in 1953, which included studies of Kaduna and Zaria conducted in 1955–6. The primary purpose of these surveys was to collect information on the expenditure patterns of urban wage-earning households with the object of establishing 'weights' for the construction of a consumers' price index. The surveys therefore assessed expenditure on accommodation as a component of total household expenditure. In addition, general basic data were also collected on the household and its housing conditions.

In 1962–3, a second urban consumers' survey was carried out in Kaduna. This covered samples of both wage-earner and self-employed households in the lower income group, as well as middle-income wage-earners. The housing inquiries of the Federal Office of Statistics grew out of the urban consumers' surveys. Their purpose also included adjustment of the accommodation component of the consumer price index; but the housing inquiries represent a shift of emphasis towards the provision of comprehensive housing data as such.

The first housing inquiry was conducted in Lagos in June 1961, the results being published only four months later. Housing inquiries were executed in fifteen Nigerian towns during 1970–1; some data from these surveys are available, but no results from housing inquiries have been published since 1961. A list of tables from the 1971–2 surveys, which covered forty-one Nigerian towns, is available from Lagos. Although subsequent housing inquiries have been carried out in Kaduna, in 1973, 1974 and 1976, the raw data were all sent to Lagos, and no analyses are so far available.

Master plans are the other main source of housing data. The first master plan to be prepared for a northern Nigerian town, the Trevallion Plan for Kano, gives only some information about densities. But the Max Lock Plan for Kaduna contains the results of a sample survey of 1 in 13 compounds throughout the densely populated areas of the city. Like the Federal Office of Statistics housing inquiries, the 1965 Lock social survey excluded government quarters and institutional housing (such as military quarters). The Lock Plan also contains data from an employment survey, where the sampling frame was a directory of the larger employers in Kaduna.

As part of the Zaria urban study for the master plan, a housing survey was carried out (Zaria Urban Study 1975). This survey covered a 10 per cent random sample of households throughout the Zaria urban area. The Zaria survey is one of the most comprehensive housing surveys yet conducted in a Nigerian town.

The available housing data are concentrated on the towns. The Federal Office of Statistics housing inquiries have, since 1971, covered a considerable number of northern towns, including Kano, Sokoto, Minna, Katsina, Jos, Maiduguri and Bauchi. Full analyses of these surveys have not yet been published. Finally, the sources of housing data used in the preparation for development plans are also available. A more recent study is a set of reports prepared for the federal government in 1973 by Doxiadis Associates, on conditions and needs in twenty urban centres.

Whatever be the limitations of these surveys, three essential features emerge about the housing situation in Nigeria. First, in many towns, a high proportion of houses are built with mud; second, in urban areas, high population density coexists with poor facilities of water supply and sewage disposal; and third, low income households spend a higher proportion of their incomes on rent than is the case among middle income groups. These conclusions can also apply to almost all African countries. What we want to do is to consider what we mean by 'housing need'. Are we merely judging the Africans' conception of 'housing need' by Western standards?

NEED FOR HOUSING

Let us first discuss in some detail the concepts and considerations that go into planning for housing. Housing 'need' is a planning concept. It expresses the difference between actual conditions and defined minimum standards. The minimum standards (norms) are specified by planners. But the extent of normative need depends entirely on the nature of the standard adopted by the planners. To illustrate this point, consider the following example. Suppose that 240 persons are living in 100 rooms; if the planning norm is 2 persons per room, then there is normative need of 20 rooms to rehouse 40 persons. If the planning norm is 2·5 persons per room, then there is no need for new rooms. The lower the norm, the less the need.

Components of Housing Need

The demand for housing is clearly tied to (1) the population's size and its age – sex distribution, (2) the rate of household formation and dissolution, (3) migration and population redistribution, and (4) the level and distribution of income. Migration influences not only the statistical pattern of demand for housing but also the aggregate demand for dwelling units. Current needs are further related to two other norms: (1) the minimum quality of housing structures; and (2) future standards. Calculations of housing needs are therefore dominated by concern with population densities and the quality of housing structures.

The methods usually take three major components into consideration. First, consider needs arising from new household formation. Annual rates of population increase in African cities are typically 2·5 per cent per annum due to natural increase alone (see, for example, Population Reference Bureau, 1975), and including net migration they are 5 or 10 per cent per annum or even higher

(Seymour, 1978c). If this new urban population is to form new households at a rate that will maintain the present average household size in the city, then 'new household formation' will be the predominant element in housing needs. Future increases in total population are difficult to project because of the dearth of reliable statistics in Africa.

Second, consider needs arising from presently overcrowded conditions. The extent of such needs depends upon the planning standard chosen for 'socially acceptable' density, usually expressed in terms of occupancy rates (persons per room). Calculations of this component of need must also specify whether the planning objective is a gradual reduction or complete elimination of overcrowding.

Third, consider needs arising from substandard conditions in the present housing stock. Planning standards must be specified with respect to the meanings of 'obsolete' and 'insanitary' dwellings. The need for new housing units depends also on the extent to which the present substandard houses can be improved or, alternatively, must be replaced. An example is given from a Zaria urban study of the percentages of components falling into each of these three categories.

Table 5.1 *Estimated housing needs of Zaria urban area for the period 1975-80.*

Component of housing need	*Number of units needed*	*Component as % of total needs*
New household formation	5,864	77.8
Relief of overcrowding	1,115	14.8
Replacement of obsolescent stock	556	7.4
Total needs	7,535	100.0

Source: Zaria Urban Study, Zaria Housing Survey (1975).

In the example given in Table 5.1, two methods can be used for calculating new household formation.

Method 1

This method uses the population growth rate (r and the base population (P_t) to project the future total population size (P_{t+n}). The current average household size (h) is then used to estimate the new household formation (W). Now,

$$P_{t+n} = P_t e^{rt}$$

where P_t is the population at time t; P_{t+n} is the population at time $(t+n)$; and t is time in years. We can estimate new household formation as

$$W = \frac{P_{t+n}}{h}$$

Using data from the Zaria urban survey, we have

$$h = 7{\cdot}2r = 4{\cdot}3\% \text{ p.a.}$$

$$P_t = P_{1975} = 176{,}000;\ t = 5 \text{ years}$$

Therefore,

$$P_{t+n} = P_{1980} = 176{,}000\ (e^{0\cdot043 \times 5})$$
$$= 176{,}000\ (e^{0\cdot215})$$
$$= 176{,}000 \times (1{\cdot}2399)$$
$$= 218{,}222$$

$$\text{New population } 1975-80 = 218{,}222 - 176{,}000$$
$$= 42{,}222$$

Hence,

$$W = \frac{P_{t+n}}{h} = \frac{42{,}222}{7{\cdot}2}$$
$$= 5{,}864 \text{ (see Table 5.1).}$$

The method has one serious limitation. In Africa, there is a great number of 'potential' or 'concealed' households – that is, groups of persons who have not yet constituted a separate household, but who would do so if they could obtain a dwelling – for example, married couples living with the parents of one of the spouses. Under conditions of very rapid urban growth and severe housing shortage, the average size of households is likely to be greatly inflated by the absorption of 'potential' households.

'Extended family' obligations promote the fusion of potentially separate households within a single shared dwelling. This creates important problems for those who attempt to calculate housing needs. Any estimates of the extent to which these relatives were sharing voluntarily would inevitably be arbitrary and subjective. And yet it is logical that present 'potential' households must be equally in 'need' as those anticipated for the future.

Method 2

The second method, called the delta method, has been proposed by the United Nations. The statistic for delta (Δ) is defined as

$$\Delta = \frac{1}{k}\left(\frac{p}{h_3}\right)$$

where p is the population, h_3 the average household size, and k the number of 'households' occupying a single housing unit.

Example: Let p (the population) be 1 million, and h_3 (the average household size) be 5. For $k = 1$, Δ (the number of housing units needed) = 200,000. With the same figures and for $k = 1{\cdot}5$, $\Delta = 133{,}332$.

Clearly, as the official norm is lowered, implying the acceptance of a greater degree of sharing, the 'need' for housing shrinks. By manipulating the norm in this way, needs arising from household formation can be made to disappear completely; those arising from high densities are not considered; and the only needs remaining are those due to substandard physical conditions.

There are two basic problems with methods for estimating demand for housing: first, the problem of forecasting marriage and divorce rates – there is no adequate data on this in Africa, as marriages and divorces are not usually registered; and second, the problem at the aggregate level in translating an age distribution such as is available from a census into a matrix of household sizes, which is then used for projecting future needs for dwelling units of different sizes. Current social trends involving the gradual replacement of extended formation by nuclear families in developing countries will raise the demand for dwelling units over time.

Overcrowding

'Overcrowding' is usually measured in terms of the average number of persons per habitable room. This also expresses the relationship between the average size of a household and the number of rooms in its dwelling. For example, if the average size of household is 3·6 and the average number of rooms per dwelling is 1·5, then there will be 2·0 persons per room. This is usually specified as an 'official room'[3] for planning purposes. The acceptable occupancy rate varies from one country to another; what is considered 'crowded' in Europe or North America may be a normal (or better than a normal) condition in most African countries. The official standard for overcrowding also varies between African countries.

This indicator of 'overcrowding' has a number of limitations. First, the ratio ignores the area and volume of the room. Second, it fails to distinguish between adult occupants and children; it is more logical to count a child as only 'half a person' in calculations of such occupancy rate. Third, outdoor living space (such as the courtyard or verandah) is important, but is excluded from the persons-per-room indicator. Fourth, when overcrowding is considered as a health problem (as a condition that contributes to the communication of respiratory infections), the relevant indicator is *per capita sleeping floor space* (in combination with an assessment of ventilation). Fifth, usefulness of the persons-per-room ratio as a measure of overcrowding is reduced when it is abstracted from overall residential density, expressed as the number of persons per hectare. Sixth, the problem in assessing 'overcrowding' concerns the choice of planning standard; a density that is desirable on human grounds may produce needs figures that it would be totally beyond the country's resources to provide. Seventh, the usefulness of the overcrowding indicator has a built-in 'Western bias' in terms of its assumptions about perception of overcrowding as a 'problem'; that is, the indicator may imply an inappropriate imposition of Western notions regarding the importance of privacy. In African cities, cultural and climatic conditions may make high occupancy rates quite acceptable. Eight, the extended family structure in Africa, with its arrangements for living together in large households, is ignored.

The calculation of 'housing' needs arising from 'overcrowding' is therefore limited in two main ways. First, on purely technical grounds, there are limitations about the persons-per-room indicator. Second, fundamentally, the calculation involves a bias in housing policy and planning towards provision of more housing units rather than improvement of the urban environment.

Substandard Housing Stock

Present 'replacement needs' are the number of units in the housing stock that are deemed to be 'substandard', and for which improvement is not considered feasible. Future replacement needs are calculated by estimating the expected life span of existing houses, and the dates at which replacement will become necessary.

Calculation of replacement needs is a complex and difficult task or undertaking. It requires detailed information about the housing stock, which is unlikely to be available unless a special survey is conducted. Such a survey, which requires evaluation of physical conditions, is technical in nature and beyond the competence of ordinary census enumerators. The definition of what is 'substandard' is not a simple matter, whether it refers to 'obsolete' buildings or to 'insanitary' amenities. Nor is it easy to assess objectively whether improvement (rather than replacement) would be possible.

The method of assessing the quality of the present stock has been well summarised and documented (Zaria Urban Study, n.d.). One point needs to be stressed. The methodology used (Zaria Urban Study, n.d.) to assess replacement needs is exceptional in its sophistication. Even so, the calculations of obsolescence are very approximate and tentative. There is inevitably much arbitrariness both in the assignment of various weights and scores and in the estimation of life span for each resulting points range.

These problems are compounded when calculations of *future* needs are required. The life span of existing buildings will be determined partly by the type of maintenance that its owners will apply; and this in turn may be related to financial ability and type of occupancy. Future conditions will also be affected by the extent to which environmental improvements are executed by government. Finally, in calculating future replacement needs there is a danger that the standards applied will produce needs figures greatly in excess of 'practical possibilities'. This is particularly the case in African countries.

USES OF HOUSING STATISTICS

Housing statistics furnish a systematic account of changes in the supply of housing available to the population as well as the natural, economic and social factors that have a bearing on these changes.

Housing statistics also serve as a principal indicator of the standard of living in a country. The types of houses and their facilities can be used to judge progress of a country. For example, comparisons of the facilities enjoyed by residents — such as sanitation, water, electricity and toilets – may be required to find out relative social and economic progress in two countries.

Housing statistics are essential for the planning and the construction of houses. In this connection, a number of African countries have embarked upon programmes for providing low cost houses for the poorer sections of the community. Nigeria is one such country. However, before any planning of such houses can be undertaken, one requires data on the existing houses and also an estimate of future requirements.

A number of problems have, however, been encountered in attempts to provide housing for the population. Housing policies formulated to provide subsidised facilities for the poorer sections of the population, including housing in the urban areas for the low income groups, have failed for three reasons. First, the housing standards usually adopted for public schemes are unrealistically high in relation to the conditions currently experienced by the urban poor. Second, unrealistic housing standards imply expensive construction costs; and even with substantial government subsidies, rents are beyond the economic means of the urban poor. In Nairobi, for example, the annual approved housing production is less than one-third of the annual demand. Nairobi City Council insists on using relatively high building standards. As a result, a large part of its population cannot afford even a minimum standard house (Mathot, 1978). Third, the urban poor are usually excluded from public housing schemes by factors other than rents: for example, the location of housing units and the implications for transport expenses and for access to places of self-employment.

Houses are an important source of ownership of wealth in African countries. Housing statistics can thus help in the estimation of the property and wealth of the nation. Similarly, they would be helpful in estimating wealth tax and tax on income from wealth. In this connection, housing policies can be used as a major instrument for income redistribution. This is because we know that, in most African countries, the urban worker spends 20 – 25 per cent and sometimes up to 40 per cent of his monthly income on rent (Seymour, 1978a; Sada, 1977). This high level of urban rent in relation to income is therefore a major factor in the redistribution of income in favour of the property-owning class and constitutes an obstacle in the attainment of a just and egalitarian society.

Housing statistics are also required for the estimation of national income, in which we can add up the value added (wages and profit) by each sector of the economy, including houses. In such cases, rent is generally imputed to owner-occupied houses.

LIMITATIONS OF HOUSING STATISTICS

The statistics on housing are among the least satisfactory available, partly because of the heterogeneous nature of the subject, and partly because responsibility for the collection of the statistics, where it exists, is fragmented among a wide variety of agencies. The statistics are therefore notable for their omissions, the irregularity of their collection, and their unknown and often highly suspect reliability.

Existing sources of data on housing conditions in Africa suffer from serious limitations. First, much of the available information is outdated. Second, limitation in the existing sources relates to their data collection procedures. In the Nigerian Federal Office of Statistics housing surveys, for example, the size of the sample was small and not representative of the entire housing stock. In addition, recent housing surveys appear to suffer from non-sampling errors; for example, in the 1971 – 2 housing inquiry in Nigeria, only 211 out of the sample of 270 were covered in Zaria, 234 out of 301 in Katsina, 212 out of 332 in Jos, and 157 out of 304 in Ilorin (Seymour, 1978b). Third, the failure to complete analysis and publish full results of surveys raises questions about the quality of data collected.

Another source of difficulty is the nature of the houses in these countries. People live in mud houses, not only in villages but also in cities. The nature of the house is such that a heavy downpour during the rainy season may demolish the house. If such houses are flooded away during the rainy season, and if, unfortunately, the time coincided with a survey period, it would be very difficult to take count of these houses. The problem is further aggravated in the rural sector by the lack of a transportation system. As a result, it becomes difficult to enumerate these houses. A further important problem in improving the reliability of statistics on housing is that the surveys are very expensive to carry out.

Conceptual and definition problems are also common. Egero (1973), writing on the difficulties of carrying out a housing survey, has remarked that

> . . . . the housing questions required a sudden shift over to the house and the focus for attention. The relation between the premises occupied by the household and the entire house was easily confused. Further, the translation of such terms as 'kitchen' or 'bath – shower' proved very difficult as did the interpretation in relation to variety of ways of living of the urban population of Tanzania. A sign of the difficulties encountered perhaps lay in the fact that no definition of these concepts was even attempted by the Census Headquarters.

International comparability can therefore be seriously limited by the use of different definitions of housing units by different countries. While similar houses can be compared, a conglomeration of different types of houses presents problems when one is comparing only the total number of houses. For example, we can compare the number of concrete houses in one country with that of another country, or the numbers of mud houses can be compared. However, stress must be laid on the fact that, while substantial variations can exist in the cost of constructing concrete houses in different countries, there is probably less variation in the cost of constructing mud houses, because the main input here is the cost of labour.

Most of the surveys also have a limited coverage, as we have noted from the data sources from a number of African countries. The surveys are also restricted to urban areas and to 'European'-type houses in the urban and rural areas. Apart from the scarcity of capital and trained personnel to carry out these surveys, one may argue in favour of limiting the scope of these surveys purely on logical grounds also. After all, what purpose will the statistics on mud houses serve, when we know that these houses are very temporary and will not last very long? Arguments in favour of counting these houses can also be advanced. These houses represent the magnitude of the problem of housing, and no planning of housing can succeed without a knowledge of the gravity of the problem. It can be argued, however, that the same information can be obtained from population censuses.

CONCLUDING REMARKS

The construction of a reliable housing series is a far cry in Africa. The housing data suffer from several weaknesses and are grossly unreliable. The planning for housing therefore has a weak statistical base in these countries. Under the prevailing circumstances, the best course would be to collect data on housing together with the census of population in these countries, and then to update the census results by using the national income statistics (contribution of houses to the national product). This procedure would minimise the cost of obtaining the housing statistics in African countries.

There is an urgent need to define our priorities in Africa with regard to housing standard. An alternative approach to housing research should involve a movement away from calculations of normative need and a shift towards the study of housing demand. This implies, in the first place, that the issue of price should be included in an assessment of the housing 'problem'. The methodology of needs produces only a general statement of the quantitative deficit at official standards; it specifies neither the costs of the proposed new units nor the incomes of the groups for whom the houses are intended. Calculations of housing needs ignore the question of distribution: that is, which specific income groups will obtain the new units, and at what prices?

The low incomes of most inhabitants make them unable to afford conventional public housing; it is therefore necessary to introduce alternative policies, such as self-help housing and programmes for progressive environmental improvement, in order to match housing development with effective demand. Research on housing should therefore go beyond analysis of existing physical conditions and household characteristics, and relate these factors more specifically to variations in occupational activity and income.

The study of housing demand should also include investigation of subjective preferences – a survey of opinions, or attitudes, towards housing development. This kind of research might contribute to bridging the gap between 'needs' as conceived externally by planners, and the actual housing preferences articulated by specific groups (particularly within the low income population).

Nevertheless, public opinion on appropriate standards should not be the sole consideration. Certain types of housing structure or densities of settlement may be acceptable to the population, but may involve health hazards of which the people are unaware. Planners have the responsibility of adopting minimum standards that are consistent with national resources and the level of economic development, and that take into account both the advice of health authorities and the expressed preferences of target populations. Research on housing demand, by providing information on economic activities and subjective housing preference, will make a useful contribution to improved planning, particularly for the lowest income groups.

NOTES: CHAPTER 5

1 See Chapter 12 for further reading on this topic.

2 For details of sources of statistics on housing in other African countries, see, for example, United Nations (1971b, 1971c). As a general rule, the statistics can be found in statistical bulletins or abstracts of Ministries of Housing and city council offices, which give building permits in many African countries.

3 The official indicator of 'overcrowding' is an official norm for the maximum number of persons per habitable room. For example, in urban Nigeria the official norm is 2·0 persons per room.

REFERENCES AND FURTHER READING

Ademu-John, C. (1974), *A Study of Housing Problems in Freetown and its Environs* (Freetown: Ministry of Housing and Country Planning).

Bedawi, H. Y. (1976), 'Variation in residential space standards in Zaria', *Savanna*, vol. 5, no. 1, pp. 75 – 8.

Campbell, E. (1977), 'Population and housing', paper presented to the First National Population Education Seminar, sponsored by United Nations Fund for Population Activities/UNESCO, Fourah Bay College, University of Sierra Leone, Freetown (14 – 22 December).

Doxiadis Associates International (1973), 'Nigeria: development problems and future needs of major urban centres', reports prepared for the Federal Ministry of Economic Development and Reconstruction, Report No. 7, and 'Existing conditions and present needs in Kaduna and Zaria', Report No. 15.

Egero, B. (1973), 'Non-sampling errors', in Egero and Henin (1973).

Egero, B. and Henin, R. A. (eds) (1973), *The Population of Tanzania: An Analysis of the 1967 Population Census*, Census Vol. 6 (Dar es Salaam: Bureau of Statistics).

Essien, E. S. (1976), 'Water supply in rural areas: an experience in Malumfashi, Kaduna state of Nigeria', paper presented at the annual conference of the Association of Community Physicians of Nigeria.

Fitchett, R. E. (1972), 'Survey of existing facilities and future measures to provide for adequate financing of housing programmes', report of the Economic Commission for Africa (ECA) mission to the Ministry of Housing and Country Planning, government of Sierra Leone, Freetown.

Ghana (1971), *Report of the Housing and Town Planning Committee,* Accra: Central Bureau of Statistics.

Hanson, D. (1969), *Preliminary Report on Housing in Sierra Leone* (Freetown: Ministry of Lands, Housing and Country Planning).

Lock, M. and Partners (1967), 'Kaduna, 1917 – 1967 – 2017: a survey and plan of the capital territory for the government of northern Nigeria', report submitted to the government of northern Nigeria, Kaduna.

Mascarenhas, A. C. (1973), 'Urban housing in mainland Tanzania', in Egero and Henin (1973), pp. 98 – 118.

Mathot, L. H. G. (1978), Letter to the editor of the *Guardian Weekly* (19 March).

Muana, J. L. K. (1975), *Sample Survey on Housing Conditions and Socio-Economic Characteristics of Middle Income Government Employees in Freetown* (Freetown: Ministry of Housing and Country Planning).

Nigeria (1966), *Report on Enquiries into the Income and Expenditure Pattern of Lower and Middle Income Households in Kaduna, 1962 – 63,* UCS/1966/3 (Lagos: Federal Office of Statistics).

Nigeria (1968), *Gross Domestic Product of Nigeria, 1958/59 – 1966/67* (Lagos: Federal Office of Statistics).

Nigeria (1975), *Third National Development Plan, 1975 – 80: North Central* Planning Office, Federal Ministry of Economic Development).

Nigeria (1975), *Third National development Plan, 1975 – 80: North Central State Programme* (Kaduna: North Central State of Nigeria).

Nigeria (undated), *Population Census of the Northern Region of Nigeria, 1952* (Lagos: Department of Statistics).

Nigeria (undated), *Urban Consumers' Surveys of Kaduna and Zaria, 1955 – 56* (Lagos: Department of Statistics).

Ogunpola, A. and Ojo, O. (1975), 'Housing as an indicator of urban poverty: the case of metropolitan Lagos', paper presented at the annual conference of the Nigerian Economic Society, Kano.

Okon, A. (1975), 'Housing and environmental sanitation in Zaria city', dissertation, Department of Community Medicine, Ahmadu Bello University, Zaria.

Patel, C. B. (1976), 'Survey on housing and socio-economic conditions of residential living area between Kingharman Road and George Brook', Ministry of Housing and Country Planning, Freetown (mimeo).

Population Reference Bureau (1975), *World Population Data Sheet* (Washington, DC).

Ruud, K. (1975), Chapter 17 of the *Annual Plan, 1975 – 76* (Freetown: Central Planning Unit, Ministry of Housing and Country Planning), pp. 170 – 4.

Sada, P. O. (1977), 'Urbanization and income distribution in Nigeria', paper presented at the annual conference of the Nigerian Economic Society.

Schwerdtfeger, F. W. (1971), 'Housing in Zaria', in P. Oliver (ed.), *Shelter in Africa* (London: Barrie & Jenkins), pp. 57–79.
Schwerdtfeger, F. W. (1972), 'Urban settlement patterns in northern Nigeria (Hausaland)', in P. J. Ucko, R. Tringham and G. W. Bimbleby (eds), *Man, Settlement and Urbanism* (London: Duckworth), pp. 547–56.
Schwerdtfeger, F. W. (1975), 'Comparative study of conventional urban houses in three regions in Africa', Ph.D. thesis, University of London, London.
Seymour, T. (1978a), 'Housing policy and income distribution in Kaduna', seminar paper presented to the Social Science staff, Ahmadu Bello University, Zaria.
Seymour, T. (1978b), 'Housing conditions in towns of northern Nigeria: a review of existing data', Centre for Social and Economic Research, Ahmadu Bello University, Zaria (mimeo).
Seymour, T. (1978c), 'Housing: a review of methodology', Centre for Social and Economic Research, Ahmadu Bello University, Zaria (mimeo).
Thomas, A. C. (1971), *Kissy Low Cost Housing: Report of a Socio-Economic Survey* (Freetown: Ministry of Housing and Country Planning).
Thomas, A. C. (1973), 'Population projections and estimation of housing needs for Sierra Leone', unpublished MA thesis, University of Ghana, Legon, Accra.
Thomas, A. C. (1975), 'Population growth and housing in Sierra Leone', paper presented at a conference on Implications of Development on Population, Institute of African Studies, Fourah Bay College, University of Sierra Leone, Freetown.
Trevallion, B. A. W. (1966), *Metropolitan Kano: Report on the Twenty Year Development Plan, 1963–83* (Kano: Greater Kano Planning Authority).
Tribe, M. (1972), 'Patterns of urban housing demand in Uganda', in J. Hutton (ed.), *Urban Challenge in East Africa* (Nairobi: East African Publishing House), pp. 138–56.
United Nations (1965), *World Housing Conditions and Estimated Housing Conditions* (New York: UN Department of Economic and Social Affairs).
United Nations (1971a), *World Housing Survey* (New York).
United Nations (1971b), *Handbook on Population and Housing Census Methods* (New York).
United Nations (1971c), *The Compendium of Housing Statistics* (New York).
United Nations (1973), *Integration of Housing into National Development Plans: A Systems Approach* (New York: UN Department of Economic and Social Affairs).
United Nations (1973), *Methods of Projecting Households and Families*, Manual 7, No. ST/50A/SERA/54 (New York: UN Department of Economic and Social Affairs).
United Nations (undated), *Survey of Economic Conditions in Africa, 1972 and 1973* (New York).
United Nations Economic Commission for Europe (1973), *Housing Requirements and Demand: Current Methods of Assessment and Problems of Estimation* (Geneva).
Zambia (undated), *Census of Population and Housing: Final Report*, Vols 1 and 2 (Central Statistics Office).

Zaria Urban Study (1975), 'Existing housing conditions and future housing requirements in Zaria urban area', draft report prepared by H. T. Bedawi for the Zaria Master Plan in the Department of Urban and Regional Planning, Ahmadu Bello University, on behalf of the Ministry of Lands and Survey, Kaduna.

CHAPTER 6

Criminal Statistics

The concept of crime prevention has been used, and is still being used, by society as an important means of keeping order. But society and civilisation have changed, and so have the nature and number of offences considered to be punishable by law. Consequently, crime is variously defined depending on time and location and in the context of a particular social system. Crime[1] or criminal offences in most societies can be defined broadly as any 'act or omission forbidden by law on pain of punishment' (Carvell and Swinfen, 1970).

When an offence is committed, it is necessary in the first place for the victim, a witness or someone who becomes acquainted with the event to realise that it is an offence; second, one of these people must report to the police, unless the police witness the event themselves; and finally, the police must decide to record such event as a crime. For offender statistics to be recorded, it is additionally necessary for the police to detect the offence and to decide to charge or caution the offender, and, if the case goes to court, for the offender to plead and to be found guilty or innocent. The sequence of activities described in this paragraph is reflected in what has become known as 'criminal statistics'. This chapter discusses the nature and scope of criminal statistics in Africa, factors affecting the reliability of the data, sources and uses of criminal statistics, and problems of interpretation.

NATURE AND SCOPE OF CRIME IN AFRICA

In recent years, African societies are believed to be facing increased incidence of criminality. Odekunle (1978), in a recent review[2] of crime in Nigeria, has noted that:

> Bribery, corruption, white-collar, and corporate crimes, which cost us (economically, socially and morally) more than the thousands of petty crimes for which our prisons are over-crowded, have come to be accepted by almost every Nigerian as part of our 'normal' existence. Our daily and Sunday papers are, in some respects, crime bulletins. There was the prevalence of armed robbery that forced the government to decree the death penalty for anyone found guilty of the offence.

Odekunle has concluded that:

> . . . crime problems exist in Nigeria and, had there been uniform and comprehensive reporting and recording systems, we would have had, statistically at least, alarming crime rates and frightening crime trends. The more important point here, however, is that the problem should be expected to worsen and become recalcitrant in the years ahead if the experience of developed nations whose line of development we are following, whose socio-economic order we are copying, is something to go by.

The situation in Nigeria is not very different from that of many other African countries. Two points, however, need to be noted: (1) that crime exists in African societies, and (2) that the crime rate is on the increase in Africa. These conclusions involve assumptions based on official statistics, or, strictly speaking, on police and law court statistics. However, not all punishable offences are detected, and not all detected offences are reported, and not all those reported are brought to court and tried. Of those that are known to the police, many go unresolved, and perhaps only a minority are cleared up.[3] Consequently, criminology has for some considerable time been concerned with the discrepancy between officially notified offences and real criminality. What, then, is the extent and nature of real criminality? The answer is clear: we do not know, at least not exactly. For want of better information, we have had to rely on criminal statistics whose reliability is increasingly being called into question. Of course, we now know that criminal statistics are nothing more than a means for recording criminality in that they provide guidance to the police and the courts on how to proceed. This poses the question of undetected crimes and also of the history and social control of offences that, although detected and reported, are not brought to trial.

In Africa, the dark figure of unreported crime is likely to be far larger than figures for Europe and America, because of geographical, police and legal factors and the preference of most people for criminal offences to be settled by traditional methods. Thus, crime statistics in Africa only represent a minority of offences known to the central government police and depend on the following factors.

Geography

In many parts of Africa, central-government police-stations are concentrated in the principal towns and urban areas and on the main roads. Tanner (1965a) has shown that the immediate consequence of the erection of a new police station is a substantial increase in reported crime. It is thus very difficult to judge to what extent a high offence or offender rate in a district reflects the criminality in the area as opposed to police activity.

Ratio of Police to People

While the geographical distribution of police stations is important in controlling crime, a more important factor appears to be the ratio of police to the people. In many African countries, while this ratio in urban areas can be quite high, the overall ratio for the country as a whole, as well as that for the rural areas, may be low. In 1963, the persons-per-policeman ratios in Kenya, England and Uganda were as follows: 645 for Kenya, 541 for England, 1,220 for Uganda. These ratios are even more extreme when they are considered on a district-by-district or rural – urban basis (Tanner, 1965a). Although there are some striking differences in district and regional crime rates which can be due to, or attributable to, cultural and social differences, the presence of the police to whom a report can be made personally can also be a crucial factor.

Police and Law Enforcement Agents' Discretion

Certain types of offences, particularly figures for drunkenness and motoring offences, are particularly subject to variation in police policy in detecting and charging offences, and this has a direct bearing on the figures or affects the recorded crime in this category. An excellent example of police discretion is given from Nigeria as follows:

> Mr. S. A. Raheen, sitting at an Ajegunle (Nigeria) magistrates court, has strongly criticised the police for not doing their job well. He lamented that these days, instead of apprehending highway robbers and other criminals that menace people's lives, the police continue to flood my court with cases of assaults that do not merit court action. (Nzeakah, 1978)

Corruption of the police or law enforcement agencies can also affect the use of their discretionary powers, as the following shows. A study of the records of Ibadan Central Police Station over a period of time has shown that, even when the accused is known:

> . . . there is no necessary relationship between the validity of a complaint and the likelihood of its being charged in court, it must first be sifted through the variable mesh of interests, influence, and bribery before a decision is taken as to whether a prosecutable

offence has been committed or not. (Bamishaiye, 1974)

And Chambliss (1975), analysing along the lines of Quinney's (1974) latest work on the criminality of crime control apparatus and personnel in capitalist societies, has stated that:

> . . . in Nigeria, the acceptance of bribes is blatantly public and virtually universal . . . Payment of bribes to the police is usually possible whenever an arrest is likely . . . It was said, and research bears this out, that one with money could pay to be excused from any type or amount of crime. Who then did get arrested? In general, those who lacked either the money or the political influence to fix a criminal charge.

Public and Government Perception of Crime and Attitudes

Public perceptions of, and attitudes towards, events may be of crucial significance to crime and offender rates. Indeed, in more serious crimes, public reporting is much more common than direct police intervention; for example, it is known that many offences such as pilfering at work are often not perceived as really criminal and that, even if they are, the witness may not report the crime because of his view of the police and their role or of possible embarrassment to himself or to third parties. Robbery with violence is considered a very serious crime by African countries, especially when it involves killing victims. This may force the authorities to over-react to the problem. For example, armed robbery has become a serious threat to the Nigerian society in recent years, as described in the following report:

> In the sprawling capital of Lagos, robbery, a crime now defiantly practised in broad day-light, along with a new wave of lynching, together far outstrip other acute social problems as traffic jams, shortage of accommodation, a galloping inflation and corruption.
>
> In a naked display of bravado, these men of the underworld enter banks and business houses during working hours and succeed in making away with important hauls while the public watch. Cars are snatched away from their owners in the thick of traffic and intercity travellers are arrested at gun point for their money or goods. The robbers stop at nothing they can get if they want, entering even schools and private residences. They often kill or maim their victims, sometimes even when no resistance is offered. (*West Africa* 1979)

Government reaction to such armed robbery in some African countries is decisive. For example, in Nigeria, under the Gowon regime, a decision was taken to execute publicly, and by firing squad, persons found guilty of armed robbery. In the first year or two the famous Bar Beach executions pulled crowds comparable only to those of a national soccer final. The executions continue, and seven years after they began it is generally admitted that they have not served any deterrent function. Instead, robberies have increased by leaps and bounds. (Amnesty International put the number of people condemned to death in Nigeria in the year up to August 1977 at 608, most of whom were armed robbers; *West Africa*, 1979: pp. 330-1.)

The problem of crime detection and punishment is further compounded by traditional values in Africa. Most tribal societies prefer to settle criminal offences by traditional methods; for example, in the rural areas, crimes are not usually reported to the various agents concerned with the apprehension of criminals because of traditional reasons. Crimes such as rape, adultery and petty theft are not usually reported by Africans in rural areas, since the stigma attached to such crimes makes the reporter of these crimes shameful in the eyes of the community.[4]

In East Africa for instance, according to Tanner (1965a), the absence of compensation as the primary theme of criminal justice has alienated the majority of East Africans from considering the courts as a source of justice. A typical remark on capital punishment is that 'it seems that by killing a murderer, government is making the murderer suffer for the murder. But for me, if my son has been murdered, there is nothing at all, neither government nor the murderer pays me anything.'

Effectiveness of the Police Force

The number of offences known to the police will depend also on the effectiveness of the police force. This statement from Nigeria illustrates the point:

> Though statistics are difficult to come by, few Nigerians are in doubt about the extent of the losses in lives, money and property the nation is suffering in what is almost a guerilla war. What is most urgent now, however, is how able are the forces of law and order in coping with the war, for how long will the plunder continue?
>
> The papers have called for more police alertness, updated equipment and increased police strength, all of which measures the Nigeria Police have adopted in recent years. It is, however, doubtful if the new measures to be adopted by the Lagos State Police will go far in stopping the highly professional and imaginative robbers who are ever in advance of police techniques. (*West Africa*, 1979).

While the importance of the effectiveness of the police force is a factor in containing crime, another factor that appears to be important is the personal assessment of the probability of being caught and convicted of a criminal act. An excellent example is given from Nigeria, as follows:

> Though violence is always premeditated, cleverness has often done the trick. In a simple eye-opening operation, a man wearing an army uniform boldly entered the heavily guarded compound of the Niger State House and stole the administrator's Peugeot car. Police and soldiers guarding the premises let him through thinking he was a new member of the staff, and it was only when Col. Joseph Oni called for the vehicle that it was realised what had happened. (*West Africa*, 1979)

Accessibility and Adequacy of the Legal System

Access to the law courts and delays in obtaining a ruling at the courts are also some of the factors that affect the reporting of crime, and hence its magnitude in any given society. In Africa, persons who witness criminal activities being committed are partly reluctant to report to the police because of delays in handling the case, and partly because citizens do not often report crime as a result of lack of conception of duty to the state in the Western sense. Clearly, the time taken to prosecute the suspects weakens the cases against them, and the time lag tends to erode the memory of prosecution witnesses.

Existing laws are modified by amendments and extent of legislation is constantly increasing, so that more laws have been amended or passed in East Africa in the few years of independence than during the colonial period. This increases the possibility of being prosecuted in general, as well as making selective prosecution for political and other purposes more likely (Tanner, 1965a).

Heredity

There is also the role of heredity in relation to criminal tendencies, which is important in the study of crime. Research on heredity and crime (Sellin and Wolfgang, 1964) is one of the areas in which statistics have contributed to criminology, and studies have suggested that individuals with an abnormal chromosome structure may be more prone to crime. For example, the presence of certain specific chromosomes in males has been linked to violent criminality. Certain combinations of chromosomes appear to be over-represented in convicted populations.

A related problem (it could be heredity and/or environmental) is the influence of parents on the criminal activities of their children. This is relevant in studies of juvenile delinquency. With juvenile delinquents, the analysis should examine types of families (divorced or separated families, unmarried mothers, and so on). The level of education of their parents, size of families, and occupation, income and other characteristics of the parents, may prove useful in highlighting factors that contribute to juvenile delinquency.

Police Recording of Crime

The methods used by police forces in recording criminal statistics have varied considerably over the years and from place to place. Hence, one aspect of police activity that shows most clearly their effect on the published statistics is the recording of offences. For organisational reasons, police may not officially record all offences that they know about. Alternatively, they may deliberately record everything in order to demonstrate the need for more men and resources for crime detection. There is also the problem created by the ambiguities of recording crime in given situations; for example, if one man steals five shirts from five different persons in a room, is this one offence or five?

In a number of cases, after a police investigation of a criminal case, it may be found that an offence rightly belongs to a classification other than the one with which the person is charged. The police records do not always make any corrections to the record previously made. Similarly, no adjustments are made in those cases where an offender is found guilty of an offence other than the one with which he is charged. Thus, the statistics of offences known to the police may have several cases of errors in classification for both offences. It is unlikely that such errors will cancel out, since the number of wrong classifications will vary considerably from one offence to another.

A major difficulty, which often arises, is the problem of reconciling the statistics of different government departments. There are, for example, cases where different figures have been recorded of persons condemned to death by courts and received into prison: that is, differences in the court statistics and the prison statistics. These may be due to mistakes of fact, misconception of instructions and carelessness on the part of officials. This particular type of error may be due to the fact that the greater the number of reporting agencies, the greater the variation in the recording and possible interpretation of a particular crime.

Attempts have, however, been made to introduce greater uniformity of recording between different areas, but many problems still remain. Increased accuracy of reporting usually gives rise to recorded increase, sometimes of a dramatic but purely artificial nature.

The above considerations emphasise that the published criminal statistics are the end result of a complex series of perceptions, definitions and decisions. Recent work has stressed the need to study this process itself in sociological terms rather than to see the limitations of criminal statistics as technical blemishes. Interpreting statistics in the former light may prove a more fruitful path for research.

MAIN SOURCES OF CRIMINAL STATISTICS

There are two main sources of criminal statistics: official records and surveys.

Official Records

The first major source of criminal statistics is the records of the police at the district, regional and main offices in the country. It is here that one obtains the statistics of criminal offences known to the police. However, the number of crimes in any community can never really be known, since a vast number of crimes are committed but not reported by the public, nor discovered by the police.

The law courts also provide a fruitful source of statistics for the outcome of persons alleged to have committed a crime. This source of crime statistics is the central government court's return of all cases registered, pending and heard, with the accused's name, age and alleged offence and the court's decision. The total number of persons dealt with in any court is described in the criminal statistics with some accuracy.

From these details, it is possible to find the proportions prosecuted on any type of charge and found guilty and the range of sentences related to the place and time of prosecution and to a particular magistrate. A major difficulty in using this source of statistics is that cases coming to court are a sample of police-selected cases, depending on the nature of the crime and the availability of the necessary evidence.

The native authority and local government authority[5] in some African countries do try suspected criminals, but records are usually not very reliable, if they are kept at all. This is another source of criminal statistics.

Survey Methods of Collecting Data on Crime

(1) *Victim Studies*

This is a method used in collecting data in order to test the validity of official offence data that have been recorded. The method has been used in a number of African countries: for example, in Nigeria.[6] The method is confined to crime with victims and consists of asking samples of the general population whether they have been victims of particular offences in the preceding year, or in a stated interval of time period. Investigations of this type have shown that, at most, only about half of all serious crimes are reported, and the factor varies according to the type of crime. Two main reasons are responsible for this discrepancy: the first reason for not reporting an offence is the feeling that the police would not, or could not, do anything about the matter; and second, other respondents felt that the matter was personal and private or did not wish to harm the offender.

Methodological problems still remain, the major one being that of recall; some may forget an offence, while others may bring forward earlier events into the relevant year. The second problem is that those who do not perceive the event as really criminal at the time will, presumably, still not do so. A minor problem, which leads to underestimation of offence rates in surveys of this nature, is that persons are not always aware when other members of the household have been victims.

(2) *Self-Report Studies*

The method used to investigate unrecorded offender data is to ask people what offences they have committed. The results[7] of these studies have produced some extremely interesting findings; for example, official offender data in developed countries indicate a high prevalence of criminal activity in males, in adolescents, in urban dwellers and in those of low social class. On the whole, self-report studies strongly confirm the male – female and the urban – rural differentiation. Age rates cannot be compared, as most self-report studies have been almost exclusively on school children and young adults. There are, however, many methodological difficulties in self-report studies.

(3) *Follow-up Studies on Convicted Offenders*

Another approach in the study of crime is the follow-up study on individual convicted offenders or criminals. This approach is cumbersome, because it may become difficult to trace the offenders, especially if they change their names and addresses after they have been discharged.

AVAILABILITY OF CRIMINAL STATISTICS IN AFRICA

The coverage of statistics on crime is very uneven in Africa both within and between countries; there are also differences in the tabulation of what is presented in the national statistical abstracts, where statistics on crime are reported for most African countries.

In Nigeria, for example, the statistics on crime are available in the *Annual Abstract of Statistics* published[8] by the Federal Office of Statistics for the whole of the federation. Here one finds the statistics of persons tried in the courts by type of offence, disposal of cases, sentences imposed on summary convictions, persons in prison by sex and length of sentence, committals to prison, persons in the federal prisons, average monthly prison population (all prisons), cases of Indian hemp in federal prisons, lunatics and lepers in federal prisons, recidivism in all prisons and, finally, young persons admitted to prison.

Two main sources are used for the compilation of the statistics in the annual statistical abstracts: the High Court of Justice and the Federal Ministry of Internal Affairs, Nigerian Prisons Service, Lagos.

The information is provided in a table for each of the regions, and later (after the creation of states) for the states. The problem is that the information provided is limited to magistrate's courts and the high courts, even though cases tried in these courts form only a proportion (a negligible proportion in the northern states) of the total number of cases tried in the country. Information should be provided on the cases tried in the Sharia courts, local government courts and customary courts. Thus, the statistics on the coverage of crime are uneven within the country itself.

The published statistics on crime in Nigeria do not give any information on the strength of the police force and its distribution over the country; neither do the published statistics give information on the offences known to the police in the country. This gap in the published statistics makes it impossible to estimate the percentage of the cases known to the police that have been cleared up on an annual basis, thus offering insight into the efficiency of the police force.

The published statistics should also provide complete information on actions taken on offenders or suspected offenders: for example, the number of offenders who have been cautioned by the police, the number proceeded against at the magistrate's courts and the number tried at the high courts. The last two categories of statistics are available in the statistical abstracts, but no information on those offenders or suspected offenders who have been cautioned by the police are available.

The statistics on crime in East African countries – Kenya, Uganda and Tanzania – are published in the yearly statistical abstracts of the three countries, and the amount of information presented on each country is fairly uniform. The main sources of these statistics are the Ministries of Home Affairs, prison departments and judicial departments. The statistics usually relate to offences known to the police and cleared, the strength of the police force, the prison population and persons tried in the courts by type of offence. The data on crime and punishment in Kenya are presented for the resident magistrate's courts, the high courts and the Court of Appeal. The types of crime are divided into civil and criminal cases.

Finally, we make one suggestion for the improvement of presentation of criminal statistics in Africa. Individuals and groups give different weights to different offences. But legal systems inevitably work on the basis of a set of highly technical rules and procedures, for two reasons: in the interests of fairness and uniformity. Therefore, it may be desirable to categorise offences in some order of gravity. Motoring offences range from trivial parking offences to highly dangerous driving resulting in death. In some advanced countries a distinction between crimes and offences is made in presenting the statistics, in much the same way as a distinction is made between indictable and non-indictable offences in African countries. It may be useful to go beyond this simple dichotomy and recognise a number of categories of gravity in the presentation of crime statistics in a uniform way.[9]

USES OF CRIMINAL STATISTICS

Crime is a social problem and the concern of the entire community. Unfortunately, crime is often looked at in Africa as un-African, a result of colonialism and the injection of alien values. But this attitude can only prevent a proper understanding of the social reality that crime cannot be abolished in any society; it can only be partially controlled. Criminal statistics from official sources as well as from research findings can therefore be used for:

(1) examining the nature of social control of crime in any given society;
(2) predicting the pattern of crime in a society, because the probability of an offender's committing another offence is usually related to several factors in his previous criminal career and in his social background; and
(3) explaining the present crime pattern in order to be able to assist in the prevention of future crime. To this end, an understanding of motive is essential, although this is not usually available in police details of cases. However, some idea can be obtained from reading court judgements.

The greater is man's understanding of the reasons for crime and of the conditions that create it, the greater will be the chance of overcoming it. This excellent example from Nigeria illustrates the importance of understanding the motives behind crime and notes:

> Either in response to the heightened outcry, or simply out of conscious duty, Lagos State Police Commissioner C. Akagbosu recently announced plans to establish new police posts and several more check points in Lagos, as well as to embark upon systematic gathering of intelligence reports on criminals. (*West Africa*, 1979)

It has also been argued that inequalities in the distribution of income and other benefits of economic development can encourage increases in criminality (see, for example, Odekunle, 1978). A correspondent writing in *West Africa* about crime problems in Nigeria has agreed:

> Many have argued that the phenomenon [crime] is a fight-back at Nigeria's deeply entrenched capitalism which controls all of the country's vast oil wealth, leaving the poor rural folk still in a state of subsistence. Thus, goes the argument, if life could be made more abundant for these hapless people, the rich might be more secure in that which would be left of theirs. (*West Africa*, 1979)

The approach has to get more and more sophisticated, for man himself is a sophisticated and complicated individual. It is necessary to keep crime rates in any country to a minimum, because these have implications for foreign investment in a country and for the upkeep of law and order.

METHODS AND PROBLEMS OF ANALYSIS OF CRIMINAL STATISTICS

The role of statistics in the analysis of social events is now well known, and statistics are used in criminology, as in other social sciences, to elucidate problems related to explanation and to social policy. Although the use of appropriate statistical methods and procedures can be of considerable importance in the study of crime, the types of problems encountered in the analysis of crime and the nature of the data are such that considerable knowledge of crime is necessary in order to use statistical techniques effectively in the analysis of criminal behaviour.

First, let us describe the nature of statistics usually available for analysis. The most frequently used measure of the amount of any kind of crime dealt with by the courts is the number of persons found guilty, and this is usually given in tables that show the punishment, or treatment, awarded or other methods by which the case may have been disposed of.

However, the statistical basis of the tables relating to court proceedings needs to be described carefully, because a criminal may be proceeded against more than once in a year, may have been convicted of several different offences on each occasion, and may ask on each occasion that a number of other offences with which he has been charged shall be taken into account when sentence is passed.

It is impossible to record every aspect of this situation in ordinary tables, and sometimes only the principal offence of which the person is convicted may be recorded. On the assumption that in a given case the recorded data are useful and reasonably accurate, what methods can be used in the analysis of criminal statistics, and what are some of the limitations of these methods? Offence rates are usually given as per 100,000 of the population, and are measured as crude rates of criminality within a given society. The limitations and usefulness of crude rates in the analysis of social phenomena have already been discussed fully. In the case of criminal statistics, an age-adjusted rate or figure would be more meaningful.

One approach that can be used is to examine rates of offence with particular reference to the exposed population and the degree of risk (i.e. the probability of a residential area's being burgled). These figures are probably more meaningful than simple crude rates. It is therefore important to use some form of standardised index in making comparisons between different areas within the same country.

Index of Crime

An attempt to solve some of these problems has been made using an 'index of crime'. The index is derived by giving each known offence or component of offence a score according to the seriousness, and the total score is based on all offences in a specified period, adjusted for the relevant population. The index only includes crimes involving injury, theft or property damage.

There are several difficulties with this index. First, there is insufficient justification for pooling the sets of scores for only these three groups of crime, and the seriousness of a particular crime may have to be determined in any given society, given the legal system and the public perception of the crime. Second, the assumption of a ratio scale and the additivity of the seriousness scores of components have also not been examined carefully. The first of these points is particularly important, as there is other evidence of considerable disconsensus in evaluating behaviour in different segments of the community, and this calls into question the whole concept of a uniform seriousness index. The only warning that we can give here is that, in Africa, the degree of sophistication in the techniques employed should correspond to the strength and reliability of the available data.

Apart from the use of standardised indices of crime, the following procedures can also be used:

(1) The area of undetected and unrecorded criminality can only be outlined vaguely. This is why previous research and proposed solutions have been unsatisfactory. One reason for this was that the ratio of detected to undetected criminality was assumed to be a constant. Apparently, this basic assumption allowed the 'hidden areas' of criminality to be neglected and attention to concentrate on recorded delinquency. To the criminologist it seemed immaterial whether he surveyed crime in its totality or only a fraction of it, provided that care was taken to ensure that the sample examined was representative and that the facts were symptomatic of the totality of crime.

The volume of unreported crime depends ultimately on public feeling associated with the particular crime at any given time, the co-operation of the public, the efficiency of the police force and other factors. The variations due to changes in public opinion cannot be measured and will necessarily be included in any changes in the number of offences known to the police. The proportion of unreported crime will therefore be high for some offences and low for others; for example, manslaughter and assaults may be highly reported, whereas petty theft may not be. Accordingly, in the analysis of offences known to the police, we can regard such statistics as purely relative and consider only their upward and downward movements over time periods.

(2) It is always better to consider one offence, or a fairly homogeneous group of offences, at a time and examine the trend: for example, the classification and analysis of statistics known to the police and cleared over the years, or their relative frequencies over the years.

(3) It is generally desirable to examine all the available statistics about whatever offence is being considered.

(4) The figures for the different police districts, so far as they are available, may often be studied with advantage. If in a district, or a group of districts, the trend is different from that of the rest of the country, possible reasons for the variation and hence for the trend may suggest themselves. Rural – urban, racial and ethnic groups, and religious affiliations are important.

(5) It is also necessary to keep an eye on movements of population or

similar changes that may have their effect on the number of offences.

(6) Studies of patterns in reconviction rates and the interpretation of the statistics are difficult, but some idea of this may be obtained from the statistics of the courts, where convicted offenders are tabulated by the number of times that they have been previously convicted. In the area of penal treatment, there are considerable problems of methodology and interpretation.

PROBLEMS OF INTERPRETATION

We have stressed the point that the statistics on crime in Africa are notable for their omissions, the irregularity of their collection and their often highly-suspect reliability. Accordingly, African crime statistics, particularly in published reports, should be treated with extreme caution, except as a basis for further inquiry. The caution is necessary because of the serious problems that the police themselves have to face in recording crime, and the reluctance of the general public to report crime, particularly to sparsely-distributed central-government tribal police forces.

It is difficult to analyse criminal behaviour, or indeed criminal deviance, when many of the recorded cases in certain fields may be police instigated (Tanner, 1965a). In East Africa, for example, in disregard of any possible effects on their public relations, the police often favour drives in pursuit of the minor petty criminals connected with licences,[10] tax, transport permits and immigration. Sometimes, the police deliberately inflate the statistics in order to provide evidence in defence of their budget allocation. The police may also be unaware of the real benefits that can accrue from a more complex, accurate and expensive system of statistics.

It is therefore necessary to treat official crime data with extreme caution, not so much to dispute their accuracy, since few police officers would suggest that they are accurate, as to regard their manifest inaccuracy as an indication of problems rather than statements of exact facts. For example, Tanner has noted that the wide variation in homicide rates found in Uganda among various ethnic groups and regions in 1963 can be regarded as the beginning of an inquiry rather than its end, particularly as the case details show the total to be made up almost entirely of violent deaths, with the almost complete absence of deaths by poisoning. From the available statistics it may be possible to conclude that poisoning does not occur in Uganda, but the infrequency of inquests, rapid burial in a hot climate and the lack of facilities for autopsies make it possible that only a minority of murders are investigated thoroughly.

The motives attributed by the police for homicide and attempted homicide for 1961 – 3 in Uganda are essentially easy explanations of immediate causes (Tanner, 1965a). Indeed, no police force has the specialist staff in Africa or the need to attempt to find the real causes of such offences, existing as they do in multiple roots of the personalities of the persons involved. The quarrel at beer bars that claimed 158 deaths or near-deaths in 1963 can hardly have been entirely fortuitous (Bohannan, 1960), just as 147 cases of wife-beating or domestic disputes give the circumstances of the offences, not the motives.

An important difficulty in the interpretation of the published figures comes about as a result of changes in legal categories of offence from time to time, making comparability of recorded rates over a period very difficult. Differences in definition between countries with different legal systems also often make useful international comparisons virtually impossible.

Finally, criminology as a subject faces exceptionally difficult problems of measurement, because, to give an adequate statement about the application of statistics to criminology, it becomes necessary to say more about criminology than about statistics. Statistics seem to be almost certainly less influential in the subject matter of criminology, due to shifts in the subject matter of criminology away from the traditional middle-range empirical studies towards concern with both the nature of crime in relation to the overall social structure of any given society, and the deviant's subjective perception of the meaning of his action. As a result of these difficulties, the published statistics are now viewed much more critically than before.

CONCLUDING REMARKS

A feature of recent criminology has been its growing closeness to the sociology of law. Methodological difficulties of victim and self-report studies all illustrate this point, as they lead naturally to a consideration of attitudes to, and knowledge of, law, but so far there have been few opportunities to study the effect of changes of the law on attitudes and behaviour. It has been noted that the official statistics on crime never reveal anything resembling real trends or distributions, but this may be an overexaggeration of problems inherent in the statistics, since victim and self-report studies have shown, in some cases, real distributions of crimes and offenders. The truth is more complex. Criminal statistics do reflect real distributions, but these can never be ascertained simply from the statistics themselves; hence, research studies 'to go behind' the official figures must be an essential component of any modern criminological work, which at any stage requires reasonably accurate data on crimes or offenders, and suitable statistical analysis of the data.

Although statistical methods have proved useful in several criminological fields that are closely related to social policy, the trend in recent years has been away from the use of statistical methods. Sociologists stress the close connection between the state and crime and the subjective–phenomenological approach to crime analysis. This approach is difficult to combine with quantification. The pendulum will probably swing back to a more positive view of statistics than prevails at present. But for this to occur, it is essential that those concerned with statistics in criminology do not repeat the mistakes of the past by tacitly ignoring the deficiencies of criminal statistics as indicators of true crime rates.

In spite of the recent trend in the approaches to crime analysis, which indicates less reliance on the use of statistics for this purpose, we end this chapter by indicating some statistical methods that have been applied in recent years to crime data.

Systems analysis has been used for studying the effect of changes in the system, such as more intensive detection, on the costs of the system, and numbers placed in institutions are important. One of the main difficulties in this type of model is the problem of assembling the data (details are needed, for each individual, of the penal record, type of offence, sentence and further offences). A measure of effectiveness is also difficult to specify, as the number of offences prevented is unknown. Multivariate analysis has also made some contribution to criminological explanation in the development of typologies of offenders, but these methods must be treated with caution, unless they are carefully related to criminological concepts. Important methodological and statistical deficiencies still remain.

NOTES: CHAPTER 6

1 Crime may also be defined as infraction of a law as stipulated by the state or government. A criminal is a person who has committed a crime and who has been apprehended, tried by a court of justice, proved guilty of commission of the offence and, finally, imprisoned or punished by the government (see, for example, Muga, 1975).

2 In a more recent article, Nigeria's crime wave was highlighted as follows: 'Popular concern about Nigeria's alarming crime rate has reached a new crescendo in recent days, with more and more editorials calling for more imaginative methods and better police equipment to cope with the country's highly professional army of robbers' (*West Africa*, 1979).

3 The Uganda police in 1963 were able to prosecute 33 per cent of penal code cases reported to them, and this included cases in which the accused was found not guilty and discharged (Tanner, 1965a).

4 See a discussion on this in Muga (1975, p. viii).

5 Until 1967, when the police forces were unified under the Nigerian Police Force (NPF), Nigeria had three police organisations.

6 These methods were used mostly in the developed countries, but there are a few examples of their use in a number of African countries (e.g. Odekunle, 1976b).

7 In a series of informal inquiries with students of Makerere University, Kampala, Uganda, none denied that they had stolen, and a majority admitted that they had taken marijuana. The majority of such admitted offences are not reported to the police, partly because of the inability of anyone to comprehend the value of the probable statutory penalties (Tanner, 1965b).

8 Prison annual reports have not been published since 1965, and the Nigerian Police Force has not published its annual report since 1966.

9 The weighting problems involved in establishing an index of crime have been discussed by Sellin and Wolfgang (1964).

10 In 1964, the case list for traffic offences for Mombasa shows that the monthly totals fluctuated wildly; July, August and September gave 485, 268 and 663 cases respectively, and then came the enormous rise to 1,732 in November. This cannot have been due to fluctuation in traffic offenders, but must have been due to police emphasis on different types of offences in different months.

REFERENCES AND FURTHER READING

Bamishaiye, A. (1974), 'The spatial distribution of juvenile delinquency and adult crime in the city of Ibadan', *International Journal of Criminology and Penology*, 2, pp. 65–83.

Beattie, R. H.(1955), 'Problems of criminal statistics in the United States', *J. Crim. L.C and Pol. Science*, 46.

Beattie, R. H. (1960), 'Criminal statistics in the United States', *J. Crim. L.C and Pol. Science*, 51.

Becker, H. S. (1963), *Outsiders: Studies in the Sociology of Deviance* (New York: Free Press).

Bohannan, P. (1960), *African Homicide and Suicide* (Princton, NJ: Princeton University Press).

Carvell, I. G. and Swinfen, G. E. (1970), *Criminal Law and Procedure* (London: Sweet & Maxwell).

Chambliss, W. J. (1975), 'The political economy of crime: a comparative study of Nigeria and the USA', in Taylor, Walton and Young (eds), *Critical Criminology* (London: Routledge & Kegan Paul), p. 171.

Clairmonte, P. and Lambo, T. A. (undated), 'Attribution of stigma as factor in juvenile recidivism in Nigeria: a preliminary inquiry', paper that was to be published in the *International Review of Criminal Policy*.

Clifford, W. (1974), *An Introduction to African Criminology* (Nairobi: Oxford University Press), p. 1870.

Clinard, M. B. (undated), *The Black Market* (New York: Holt, Rinehart & Winston).

Clinard, M. B. (undated), *Crime in Developing Societies* (New York: Holt, Rinehart & Winston).

Cloward, R. A. and Ohlin, L. E. (undated), *Delinquency and Opportunity: A Theory of Delinquent Gangs* (New York: Free Press).

Cohen, A. (1966), *Deviance and Social Control* (Englewood Cliffs, NJ: Prentice-Hall), p. 35.

Cohen, A. K. (undated), *Delinquent Boys: The Culture of the Gangs* (New York: Free Press).
Connor, W. (1972), *Deviance in Soviet Society: Crime, Delinquency, and Alcoholism* (New York: Columbia University Press).
Cressey, D. R. and Ward, D. A. (1969), *Delinquency, Crime and Social Process* (New York: Harper & Row).
Elias, T. O. (ed.) (1968), *The Prison System in Nigeria*, papers presented at the national conference on the Prison System, 1 – 5 July (Lagos: University of Lagos).
Gillin, J. L. (1945), *Criminology and Penology* (New York: Appleton-Century).
Goffman, E. (1963), *Stigma* (Englewood Cliffs, NJ: Prentice-Hall).
Gold, H. and Scarpitts, F. R. (undated), *Combating Social Problems: Techniques of Intervention* (New York: Holt, Rinehart & Winston).
Goldman, N. (1963), *Differential Selection of Juvenile Offenders for Court Appearances* (New York): NCCD).
Harting, F. E., *Crime, Law and Society* (Detroit, Mich.: Wayne State University Press).
Hood, R. and Sparks, R. (1970), *Key Issues in Criminology* (New York: McGraw-Hill), ch. 1.
Jones, H. (undated), *Crime in a Changing Society* (Harmondsworth: Penguin).
Kenya (undated), *Annual Report on the Administration of Prisons in Kenya, 1965 – 1970* (Nairobi: Ministry of Home Affairs/Government Printer).
Kenya (undated), *Statistical Abstract* (Nairobi: Statistics Division).
Lemert, E. H. (undated), *Human Deviance, Social Problems and Social Control* (Englewood Cliffs, NJ: Prentice-Hall).
Matza, D. (1964), *Delinquency and Drift* (New York: Wiley).
McClintock, F. H. (1968), *Crime in England and Wales* (London: Heinemann).
Merton, R. (1970), 'Social structure and anomie', in M. E. Wolfgang *et al.* (eds), *The Sociology of Crime and Delinquency*, 2nd edn (New York: Wiley).
Milner, A. (1972), *The Nigerian Penal System* (London: Sweet & Maxwell), esp. ch. 1: 'Crime in Nigeria'.
Muga, E. (1975), *Crime and Delinquency in Kenya* (Nairobi: East African Literature Bureau).
Murdock, G. P. (1945), 'The common denominator of cultures', in R. Linton (ed.), *The Science of Man in the World Crisis* (New York: Columbia University Press), pp. 123 – 42.
Nettler, G. (1974), *Explaining Crime* (New York: McGraw-Hill).
Nigeria (1962), *Annual Report of the Nigeria Police Force, 1961* (Lagos: Federal Ministry of Information), p. 12.
Nigeria (1966), *Annual Report of the Nigeria Police Force, 1965* (Lagos: Federal Ministry of Information), p. 9.
Nigeria (1966), *Annual Report of the Federal Ministry of Labour, Social Welfare Division, 1965* (Lagos: Federal Ministry of Information), pp. 15 – 16.
Nigeria (1974), *Public Service Review Commission: Main Report* (Lagos: Federal Ministry of Information), pp. 86 – 8.

Nzeakah, G. (1978), 'Magistrate raps police', *New Nigeria* (15 July, a daily newspaper).
Odekunle, F. (1976a), 'Efficiency in the police protection aspect of social defence administration: an assessment of the Nigeria police force', paper presented at the symposium on Productivity and Efficiency in the Public Service, University of Ife (5 – 7 July), pp. 10 and 22.
Odekunle, F. (1976b), 'The victims of crime in a developing country: a Nigerian study', paper presented at the Second International Symposium of Victimology, Boston, Mass. (5 – 11 September).
Odekunle, F. (1978), 'Capitalist economy and the crime problems in Nigeria', paper presented at a seminar at the Centre for Social and Economic Research, Ahmadu Bello University, Zaria.
Okediji, F. O. and Okediji, O. O. (1968), 'The sociological aspects of prison organisation in Nigeria', in Elias (1968).
Oloyede, E. O. (undated), 'Juvenile delinquency in Nigerian law: a case for closer study', *Nigeria Lawyers Quarterly*, vol. 5, no. 1, pp. 31 – 47.
Quinney, R. (1970), *The Problem of Crime* (New York: Dodd, Mead).
Quinney, R. (1970), *The Social Reality of Crime* (Boston, Mass.: Little, Brown).
Quinney, R. (1974), *Crime Control in Capitalist Society: A Critical Philosophy of Legal Order* (Boston, Mass.: Little, Brown).
Radzinowicz, L. and King, J. F. S. (1975), 'Concepts of crime', *Times Literary Supplement* (26 September), pp. 1087 ff.
Read, J. S. (1966), 'Some legal problems in East Africa', paper presented at the conference on Penal Problems in East Africa, University College, Dar es Salaam, Tanzania.
Rigby, P. (1971), 'Criminal behaviour in Uganda', *Uganda Argus* (29 October).
Schafter, S. (1969), *Theories in Criminology* (New York: Random House).
Sellin, T. (1951), 'The significance of records of crime', *67 Law Quarterly Review*.
Sellin, T. and Wolfgang, M. E. (1964), *The Measurement of Delinquency* (New York: Wiley).
Sellin, T. and Wolfgang, M. E. (eds) (undated), *Delinquency: Selected Studies* (New York: Wiley).
Sutherland, E. H. (undated), *White Collar Crime* (New York: Holt, Rinehart & Winston).
Sutherland, E. H. and Creassey, D. R. (undated), *Criminology*, 8th edn (Philadelphia: Lippincott).
Tanner, R. E. S. (1965a), *Some Problems of East African Crime Statistics*, Paper No. 431 (Kampala: East African Institute of Social Research, Makerere University).
Tanner, R. E. S. (1965b), *The Selective Use of Legal Systems in East Africa*, Paper No. 393 (Kampala: East African Institute of Social Research, Makerere University).
Tanner, R. E. S. (1965c), 'A projection of expected crime in Buganda and Busoga for 1970', report prepared for the Commissioner of Police, Uganda (unpublished).
Tanner, R. E. S. (1966), 'Cattle theft in Musoma, Tanzania, 1958 – 59', *Tanganyika Notes and Records* (July).

Tanner, R. E. S. (undated), 'Penal practices in Africa: some restrictions on the possibility for reform', *Journal of Modern African Studies*, vol. 10, no. 3, pp. 447 – 9.

Timms, N. (undated), *A Sociological Approach to Social Problems* (London: Routledge & Kegan Paul).

Tumin, M. (1967), *Social Stratification: The Forms and Functions of Inequality* (Englewood Cliffs, NJ: Prentice-Hall).

Turner, T. (1976), 'Multi-national corporations and the instability of the Nigerian state', *Review of African Political Economy*, no. 5, pp. 63 – 79.

Uganda (annual), *Statistical Abstract* (Entebbe: Statistics Division).

Uganda Police (undated) 'Reported offences against persons: Mbale police district, 1961 and 1964', report at Police Headquarters, Kampala.

United Nations (undated), *Fifth UN Congress on the Prevention of Crime and Treatment of Offenders*, Congress Working Paper A/Conf. 56/6 (New York, pp. 68 – 72.

Vold, G. (1958), *Theoretical Criminology* (New York: Oxford University Press).

West Africa (1979), 'Nigeria's crime wave', correspondent's report, no. 3215 (26 February, weekly).

Wolfgang, M. and Ferracuti, F. (1967), *The Subculture of Violence* (London: Tavistock).

Wolfgang, M. (1968), 'The viable future of criminology', in D. Szabo (ed.), *Criminology in Action* (Montreal: University of Montreal Press), pp. 109 – 10.

CHAPTER 7

Labour Statistics

Employment is one of the most pressing social and economic problems of our time; consequently, measures of utilisation and non-utilisation of labour are of a great deal of interest to analysts and policy-makers alike. These measures are needed to answer such questions as: How serious is the employment problem in Africa? What are the levels of unemployment and underemployment? Are the levels increasing or decreasing over the years? Unfortunately, satisfactory answers to these questions cannot be given, partly because of lack of data on manpower statistics and also partly because of conceptual difficulties regarding what is really meant by an economically active population, employment, unemployment and underemployment in Africa. For example, who is a member of the labour force? What is an economic activity? How hard must one be seeking a job in order to be classified as unemployed? How many hours must someone work in order to be classified as underemployed?

DEFINITIONS, CONCEPTS AND MEASUREMENT PROBLEMS

We now turn our attention to concepts and definitions as well as the problems of measurement associated with these conceptual difficulties in applying Western-type definitions and concepts to the African environment. The difficulties can easily be understood when one considers the confusion about these issues in countries such as Sweden and Britain, which have some of the best statistics in the world (see, for example, Bondestam, 1973, p. 27).

Economically Active Population

A population can be divided into two components: economically active, and economically inactive. A person is *economically active* if he/she is either working or looking for work. On the other hand, the *economically inactive* population consists of students, people too young to work or too old to work, disabled persons, and so on. The economically inactive population will not be treated here.

The economically active population is further divided into: employers, who operate their own business and engage other people; own account workers, who operate their own business without engaging other people (this group is sometimes confused with the employers); wage-earners; unpaid workers (often family workers); and job-seekers. A person who belongs to the last group is classified as *unemployed*. If he/she belongs to one of the other groups, he/she will be classified as either *employed* (regularly working at least half time), *underemployed* (irregularly working, e.g. most daily workers, or regularly working less than half time) or *quasi-employed* (mainly own account workers in non-productive engagement, e.g. shoe-shiners, prostitutes, street vendors and car-watchers).

The problems of the reliability and of the measurement of economic activity are well known. Egero (1973) has written on Tanzania as follows:

> Questions on economic activities gave the same very unreliable results. This is evidenced by the widespread confusion as to the classification of women in 'economic activity' and 'economic status'. The gross differences in classification principles used led the census office at some stage in the processing to institute a rule recognizing adult women in rural households as economically active. Exactly how and where this rule was applied remains unknown, but it would seem clear that no value whatsoever can be attached to the resultant findings on the economic activities of women. In fact the results for men may not be so much better; as seen in the second part of this chapter (para. 2.2.8.), the only variables for which no reliable conclusions could be reached on the size of the sample errors were the two related to economic activity.

Labour Force

The labour force is usually defined as the sum of the numbers of people employed (for pay or on their own account or as unpaid family workers) or unemployed (commonly defined as people who are not at work but are seeking work or are available for work). The rate of growth of the labour force depends upon the rate of growth[1] of the population of working age and upon participation rates: that is, the proportion of people in different sex and age groups who are employed or are seeking work or are available for work.

There are difficulties in applying this definition to African countries. First, in Africa, the distinction between members of the labour force and dependents is not at all clear, particularly in the rural areas, where men, women and children above a fairly young age devote a part of their time to productive activities. Even in urban areas there are large numbers of people who carry on certain kinds of work at least sporadically, but who are also at least partially dependent on others.

Second, the number of people wanting work varies with the amount of work available. Often 'discouraged workers' (there are many in Africa) who do not bother to look for work when they think that there is no chance of finding any start to do so if the chances improve. This may lead to changes in the size of the labour force that are very hard to predict and that depend partly on policy measures taken; the number of people for whom work is needed depends partly on the number for whom it is provided.

Employment, Unemployment and Underemployment

If it is not clear from our definition from the last section who is a member of the labour force, it will be equally unclear who is employed or is unemployed, since a person can be employed or unemployed only if he/she is a member of the labour force. From the point of view of a working definition, employment and unemployment are easy to define.

Employment is the numbers of people who work for pay or on their own account or are unpaid family workers. But is is important to stress that employment is difficult to measure in Africa on account of the sharing of the work among family members and the great variations in intensity of work at different seasons of the year (more about this later).

Unemployment is also easy to define. The figures should include all those out of work, able to work and looking for a job through official channels or by other means. The simplicity of this definition, however, presents some problems of measurement. Should the figure include,[2] for example:

(1) discouraged workers who have given up active job-hunting because there is too little work in their area?
(2) those laid off, temporarily or indefinitely? and
(3) those with a job to go to sometime in the future?

Dey and Mogil (1973) have emphasised the conceptual and measurement problems in Tanzania as follows:

> In addition to the limitations presented by arbitrary classification of the population, there is the more general problem of applying a set of concepts more applicable to a developed country than to a subsistence economy such as in Tanzania. One concept that is used inadequately is that of 'employment'. The terms 'employed' and

'working' are often more applicable to an industrially advanced nation that to one with a large subsistence economy. The term 'employed' was ambiguous to many in the census partly from the fact that the first alternative to 'employed' in the questionnaire was 'unemployed'.

The wide divergence between the unemployment rates in rural and urban areas testifies to the difficulties in making the distinction between 'employed' and 'unemployed'. The term 'employed' could easily be associated with wage employment, which would apply to a large sector of the urban situation. Hence, the distinction between 'employed' and 'unemployed' is more appropriate in urban areas than in rural areas where wage employment is a very small portion of the total employment. The difficulty in classifying workers becomes very apparent when considering casual workers, again a large number in the urban areas.

A related problem is the interpretation of the questionnaire. For example, there is no direct Swahili translation of 'own-account worker' or of 'family worker'. In dealing with the occupational and industrial classification, the enumerators were faced with a similar problem since there is no suitable translation to distinguish 'occupation' from 'industry'.

Thus far, our discussions show that, in the field of employment and unemployment, definitions must be culturally specific.

The concept of *underemployment* is not so easy to define. We shall therefore explain the term very carefully. In Africa, large numbers of people cannot be regarded as either fully employed or wholly unemployed. This has led to widespread use of the concept of underemployment. Underemployment can be classified into (1) visible underemployment (characterised by involuntary unemployment of people working shorter hours than they desire), and (2) invisible underemployment (consisting of those whose productivity is abnormally low or whose jobs do not fully utilise their skills).

The number of hours worked may be either the number of hours of work desired at prevailing rates of remuneration (in which case some people working very long hours may be considered visibly underemployed) or a number of hours considered normal or appropriate: for example, 40 – 50 hours per week. But in Africa, many people attending places of work for a full number of hours may still be said to be visibly underemployed either if they are idle for much of the time or if, although they are continuously occupied, the product of their activity is of unacceptably low value.

The concept of invisible underemployment has been further clarified by noting that people are invisibly underemployed if more of them are engaged in a certain activity than are needed to produce the output that they do produce. In this connection, we must stress that African economies are largely agrarian economies. The whole family works on the same farm, and the output produced represents a mixed effort of all the members of the family. In general, more hands are needed on the farm than are employed. If some of these persons are siphoned off to some other work, the output may not necessarily suffer. In technical terms, the marginal product on the farms is by and large zero, and therefore disguised unemployment exists in these societies. For example, if there are ten members of the family and all are working to produce x level of output, whereas only eight persons are sufficient to produce the same amount, then two persons are disguisedly unemployed. Since all the members of the family are working on the farm, it is not possible to say which person is unemployed. The magnitude of the disguised unemployment in African economies is therefore difficult to ascertain.

An important conceptual problem, however, arises in applying the definition. Clearly, the number of people needed to produce a given output will depend on: levels of nutrition, techniques of production, size of landholdings and a whole lot of other factors, which change over time and in which changes can deliberately be induced by means of appropriate policies. Because of these imprecise definitions and conceptual difficulties, there is always controversy about the level of a country's reserve of underemployed labour. A 'labour reserve cannot be defined without specifying a time span during which definite policy measures are applied to a concrete situation' (Myrdal, 1968).

Further attempts at clarifying concepts and definitions have been suggested (Myrdal, 1968). The approach is to distinguish between three dimensions in the concept of underutilisation of labour: the numbers at work, the duration of work and the intensity of work dimensions. The intensity of work seems scarcely measurable; but the productivity of, or income yielded by, work can be taken as a rough proxy, and surveys of family incomes or expenditure can be designed to throw light on this.

The point that needs emphasis is that underemployment can scarcely be regarded as a serious problem if it yields a sufficient income; one might therefore try to cut through conceptual difficulties by thinking in terms not so much of underemployment as of low productivity or low income employment. These groups of unduly low income, plus the unemployed, can be regarded as reflecting the numbers of those for whom work, or more work, needs to be provided.

Despite the various attempts to define visible underemployment and clarify conceptual issues, problems of measurement still remain. This excellent example from Ghana illustrates the point:

> There has so far been only one serious attempt at measuring the level of visible underemployment in Ghana. This was during the 1960 Post-Enumeration Survey, when a specific question – 'number of days worked during the past 4 weeks' – was asked. It may be mentioned that the results give only a rough picture of the extent of visible underemployment. For example, a person who worked one hour each day of the 4-week period would be classified as having worked 28 days, whereas a person who worked 12 hours each day for 5 days would be classified as having worked 5 days. A much better approach would have been to use the number of hours worked during the reference period. (de Graft-Johnson, 1970)

SOURCES OF LABOUR STATISTICS

National Statistical Sources

The major sources of labour statistics are many and varied. These sources range from population censuses, labour force surveys, establishment surveys, employment exchanges, and agricultural, trade and commerce censuses to social security statistics. Between these sources they cover employment, unemployment, the labour force and manpower statistics. Underemployment statistics are usually covered in sample surveys. A brief general discussion of these sources and their limitations is presented in this section.

(1) *The Population Census*
This is the most unique source of information in that it gives a complete count of all individuals. Its main weaknesses are that: (a) it is done at one point in time; (b) it is very sensitive to changes in definitions; (c) in African countries, where employment is seasonal, this results in undercounts; and (d) its analytical usefulness is limited by the fact that it provides less detail. All these limitations notwithstanding, the census remains a unique source of information.

(2) *Labour-Force Sample Surveys*
Household sample surveys are a source of regular information on the total civilian labour force, including employers, unpaid family workers, seasonal workers and workers on own account – groups that are not covered by the other types of employment statistics. These surveys are the best source of labour statistics. They are comprehensive and can control details missed in censuses. The questionnaire for such surveys can be tested in pilot surveys. The major limitations include definitional problems, and the

interpretation and perception of questionnaires by respondents as well as interviewers.

(3) *Establishment Surveys*

These are a good source for employment data, but are limited to the modern sector and sometimes even to the manufacturing activities only. Here the major limitation is a lack of coverage of unemployment and underemployment.

(4) *Employment Exchanges*

These could fill in the gaps in establishment surveys, as they provide data on unemployment. Their coverage is nevertheless unsatisfactory for the following reasons: (a) many people refuse to register at the exchanges; (b) some unemployed persons may register at more than one exchange; and (c) even employed persons may register with the exchanges if they are in the process of changing jobs.

(5) *Social Security Statistics*

These, like establishment surveys, cover only employment and are further limited to certain types of employment. They may cover as low as 10 per cent of wage employment. The statistics relate to the working population (employed persons covered by sickness, accident or unemployment insurance or the like). The number of contributions paid provides a measure of the insured persons in employment (unemployed persons being exempted from paying contributions). The provisions regulating an insurance scheme determine the scope of the statistics derived from its operation.

(6) *Labour Registration Statistics*

These statistics are based on the compulsory registration of persons in employment, usually at monthly or quarterly intervals. They provide a continuous census of persons in employment falling within the scope of the regulations regarding registration.

(7) *Agricultural, Trade and Commerce Censuses*

These provide supplementary data for labour statistics, but are limited by the fact that they are taken once in five years and are bedevilled by definitional problems.

(8) *Compulsory Unemployment-Insurance Statistics*

These statistics have a broad industrial coverage, which varies with national insurance legislation.

(9) *Unemployment Relief Statistics*

These show the number of persons who receive unemployment benefit from special relief funds usually administered by employment offices. The scope of these statistics is often limited, because benefits are generally small and are granted subject to many restrictions.

(10) *Statistics of Trade-Union Benefit Funds*

These statistics cover only unemployed trade-union members; their reliability depends on the stage of development of trade unions in the economy and the rules governing trade union membership or union benefits. In Africa, the statistics from this source are unreliable.

There are several problems in using population censuses and labour force surveys for obtaining labour statistics in Africa. First, the concepts of self-employment, underemployment and unemployment represent a continuum difficult to delineate. In theory, income, productivity and time worked can be the yardsticks for distinguishing between the above concepts, but in practice difficulties arise when using any of these. For example, in censuses, time worked, which varies considerably, is used to measure unemployment, while only short reference periods are used for income. Thus, there is a trade-off between long reference period and short reference period.

In Ghana, for example, there are two major problems in the labour statistics, which represent establishment surveys; these are: (1) the problem of limited coverage and non-response by establishment; and (2) the content of the questionnaire and its interpretation by respondents. The labour statistics in Ghana cover the wage employment sector and thus exclude self-employed persons. But even within wage employment, employers are excluded. The coverage of the public sector is generally better than for the private sector. There is a serious problem of non-response from private establishments. To obviate this the Bureau of Statistics has used dummies: that is, assumed that firms that had reported in a previous year would still have the same number of employees. But persistent checks have revealed that some of the establishments represented by dummies were ghost establishments.

The questionnaire for the labour statistics had items on (1) number of hours worked, (2) wages, and (3) distribution of employees by occupation and industry. The respondents always recorded official number of hours and even failed to reflect overtime worked or the time lost through casual leave and regular leave. The questionnaire limited itself to the minimum wage as the last income bracket. Thus, even though it is known that some workers earn less than the minimum wage, the reported figures in the labour statistics indicate that all workers earn at least the minimum wage.

Collection of Employment and Unemployment Data

The methods of collecting employment and unemployment data and the practices in individual countries vary from one country to another. In some African countries, a combination of methods has been adopted at the same time. First, let us look at country practices in collecting employment data. No attempt is made here to give a detailed description of each country's practices, since this does not appear to be of vital interest and since the large number of states and territories (over forty-eight) poses something of a problem. Instead, we deal with a few selected country practices to show the variations and similarities.

(1) *Country Practices in Collecting Employment Data*

In Gabon, the employment statistics are derived from compulsory social insurance, although prior to 1964 the series related to the collection of such statistics from establishments. In Ghana, the series are related to the statistics collected from establishments, and the survey is undertaken during the month of December each year. In Kenya, it is based on the statistics of establishments and covers wage-earners and salaried employees of both sexes, including apprentices and part time workers. Data are derived from the 'annual enumeration of employees and self-employed persons', conducted at the end of June each year.

In Malawi, the source of data is the statistics of establishments, which are published in the *Quarterly Digest of Statistics* by the Ministry of Finance. In Nigeria, the source of data is also based on statistics of establishments, collected annually by the Employment Service Division of the National Ministry of Labour. The survey is undertaken in the month of December each year. Prior to 1961 it used to be undertaken in the month of September. Data are collected by mail from government and private units employing at least ten or more employees throughout Nigeria. The reporting units consist of both establishments and enterprises for the private sector and for subministry, departments, councils or corporations for the public sector.

For Sierra Leone, the source data are based on statistics of establishments. Wage-earners and salaried employees (including home workers), employers and paid family workers who have worked for pay for at least one day during the month of reference are all included. However, persons absent from work during the entire period of reference because of temporary lay-off, industrial disputes or unauthorised absence are excluded. The annual data refer to the month of December.

In Zambia, the source data are the statistics of establishments and computations of data based on the last payday of each quarter. Wage-earners and salaried employees on the payrolls of reporting establishments who have worked for at least 20 hours per week for any employer are included.

(2) *Country Practices in Collecting Unemployment Data*

The practices followed in individual African countries with regard to the collection of the statistics of unemployment also vary. In the case of the Cameroun, the source is based on the employment office statistics and refers to unemployed persons registered in Douala and Yaounde. In Chad, it is based on statistics obtained at employment offices and refers to applicants looking for work. This is also true of Congo,where the data are based on the employment office statistics and refer to applicants looking for work in Brazzaville or Pointe-Noire. In Ghana, the sources of data include employment office statistics. Persons registered as unemployed and seeking for work at employment offices include workers temporarily laid off, persons seeking their first job and those seeking to re-enter the labour force after a period of voluntary inactivity. The data relate to unemployed persons registered at the end of each month. Cards for persons who have not renewed their applications are removed from the live register after four weeks.

In Guinea, information is based on employment office statistics and refers to applicants looking for work in Conakry region. The same applies to Ivory Coast, Mali, Mauritius, Togo, Niger and a few other countries. In Morocco, the series is based on the employment office statistics and refers to applicants seeking employment. Prior to 1963, the series excluded unemployed persons who did not renew their registration after thirty days.

In Mozambique, the series is based on employment office statistics and refers to the unemployed persons registered during December each year. In Nigeria the employment office statistics refer to the persons registered on the last day of the month. Persons are classified as unemployed as long as they keep renewing their registration at an employment office. The registration cards of persons who have found work are removed from the live register. Registration is voluntary.

In Sierra Leone, the series is based on employment office records and on maritime pool records. For the former, persons of 12 years of age and over (excluding seamen) are registered as unemployed and seeking work at employment offices. The data include persons seeking their first job, those temporarily laid off and persons seeking to re-enter the labour force after a period of voluntary inactivity; employed persons registered at employment offices and seeking more regular and more remunerative employment are counted as unemployed. Monthly data are computed on the basis of the live register figures for the end of the last complete calendar week of the month. Registration cards of persons who have found work are removed from the employment register, as well as cards of those who fail to report to the employment office at least once a month.

In the case of the pool records, applicants aged 12 years and above are registered at the maritime pool while awaiting sea-going employment. This register covers all persons registered in the maritime pool, including those who, after having returned from sea, must remain ashore for a period of three months before being eligible for another voyage; registered persons who remain unemployed for more than twelve months are excluded from the maritime pool records. Details can be obtained from the *Sierra Leone Gazette*.

As a general rule, the resulting statistics from all these sources are published in special census or survey volumes, sample survey reports, reports from labour exchanges, or statistical abstracts or bulletins in various countries. The frequency of publication also varies considerably, but monthly, quarterly, biannual or annual reports are generally issued.

International Statistical Sources

There are three main international sources.

(1) *Bulletin of Labour Statistics*

The *Bulletin of Labour Statistics*, published quarterly by the International Labour Organisation (ILO), presents data on employment, unemployment, hours of work, wages and consumer prices. The series published in the *Bulletin* also appears in the ILO's *Yearbook of Labour Statistics*, which in addition contains series for countries whose data do not lend themselves to publication in the *Bulletin*.

(2) *Yearbook of Labour Statistics*

This is one of the world's leading statistical reference works of source material. It brings together in a systematic, comparable form a mass of data from a vast network of authoritative sources of information in some 190 countries. The 1977 edition is the thirty-seventh volume and presents information for the past ten years. It gives details of exactly how many people did a particular job for how many hours and for how much remuneration. As well as tables giving more detailed information by industry, occupation, sex, and so on, the *Yearbook* contains data on the structure of the total and economically active population, labour productivity, industrial accidents and industrial disputes.

(3) *FAO Reports*
Other main sources of data on employment and unemployment are the reports published by the Food and Agriculture Organisation (FAO) of the United Nations, based in Rome.

Treatment of Statistics

The statistical treatment of the data is simple. Annual figures are computed as unweighted arithmetic averages of monthly or quarterly figures. Unemployment percentages are calculated by relating the number of unemployed to the civilian labour force (employed or unemployed) times 100. Whereas the absolute numbers indicate the volume of unemployment within the fields covered by the series, the percentages illustrate the severity of unemployment: that is, the relative importance of unemployment within the scope of the series. The unemployed do not form a homogeneous group; the chances of speedy re-employment or employment for the young, healthy and skilled are very much higher than for the old and unskilled. Therefore, apart from the overall unemployment rates, rates for fairly homogeneous groups should also be calculated separately.

PROBLEMS OF COMPARABILITY OF LABOUR FORCE MEASURES

According to internationally recommended standards for population censuses, the economically active population (labour force, in the terminology used for convenience in this study) is defined as those individuals who furnish the supply of labour for production of economic goods and services (United Nations, *Principles and Methods for the 1970 Population Censuses*, pp. 60–3). The concept of economic goods and services in this context corresponds to the concept of income in statistics of national accounts. Thus, the members of the labour force are the producers of a nation's income, and the remainder of the population can be considered as dependents in the sense that they consume income without taking part in the work of producing it. Included in the labour force are paid employees, employers and self-employed persons who work for profit, and unpaid family workers (relatives who assist without pay in a family-operated income-producing enterprise, such as a farm, store or handicraft industry). Unemployed workers are included, as well as those actually employed in income-producing jobs at any given time. Outside the labour force are housewives, students, retired and disabled workers, institutional inmates, young children and others who do not work at income-producing jobs, although they may receive income in the form of rents, dividends, pensions, and so on.

While this basic concept is fairly well established in census practices in most countries, there are important variations in details of the definitions and ways of formulating the census questions, which detract greatly from international comparability of the measures. Comparability is impaired further by varying interpretations that fieldworkers and respondents may give to similar questions and definitions, and by errors and biases in responses and coverage of the enumerations.

The most important point of divergence in definitions is in the classification of individuals who play a dual role, both as income-producers and as housewives, students, and so on. In some censuses, the questions refer to the individual's principal activity. Numerous part-time, seasonal and other irregular workers may then be left out of the count of the labour force. In other censuses, it is provided in principle that all persons engaged to any extent in work for pay or profit or as unpaid family workers should be counted in the labour force, although it is unlikely that a complete enumeration would be achieved in any case. The latter basis of enumeration is the one specified in international standards for population census statistics, except that in the case of unpaid family workers some minimal amount or regularity of involvement in the work of the family-operated enterprise is recommended as a qualification for their inclusion in the labour force.

Another point of divergence is the time reference of census questions about activities. Formerly, in most censuses, no particular time reference was specified; the questions referred more or less vaguely to the individual's usual occupation. The practice of asking about activities during a specified period of time has been gaining vogue in recent censuses, and this is recommended in the international standards, although some experts express doubts about the suitability of this procedure, especially for censuses in less developed countries where measures of seasonal variations are lacking. The time references specified vary; sometimes it is the census day or a brief period such as one week, sometimes several months or a year, sometimes the 'working season'. These variations may have important effects on the enumeration of seasonal and casual workers and of persons having recently joined or withdrawn from the labour force. The longer the time reference, the larger will be the measure of the labour force if all persons involved in income-producing work to any extent during the specified period are included.

Definitions of unemployed workers also vary. When the concept of usual activity was the basis of labour force enumeration, the unemployed could be identified as persons having a usual gainful occupation who were out of a job at the time of enumeration, although the measures of unemployment obtained in this way were generally not very satisfactory. In recent censuses where a specific time reference has been adopted, the unemployed have been identified by various kinds of questions relevant to availability for employment: whether they were seeking work, wanted work, and so forth, during the specified period. Variations in the forms of such questions and in details of the definitions and instructions may greatly affect census measures of unemployment, and effects on the measures of the labour force may be substantial where there is much unemployment.

Although it is in keeping with the basic concept of the labour force to include armed forces, since they are paid employees (although their service may be involuntary), armed forces are classified in some censuses with the population not in the labour force, while in others they are excluded from the coverage of the census. Sometimes a distinction is drawn in these respects between regular members of the armed forces and temporary conscripts, or between those living in military quarters and those living outside. When armed forces are either counted as not in the labour force or excluded from the census, the effect is to understate both the absolute size of the labour force and its proportion to the population. Measures of the male labour force are affected proportionately more than those of the total of both sexes, and the effect may become very important when measures for military age groups of the male population are considered.

In addition to the formal definitions, census measures of the labour force are influenced by the phrasing of the questions and arrangement of the census questionnaires, details of the instructions, and the care taken by respondents and interviewers to provide complete and accurate information. Popular preconceptions and the level of literacy in the population are also influential factors. The same questions, definitions and procedures for enumeration may produce different results in different cultural settings.

Varying definitions and other factors of non-comparability and error are likely, in general, to affect measures of the female labour force more than those of the male. Agricultural workers and unpaid family workers are groups that are most sensitive to these factors. Measures of the labour force in the rural population are likely to be affected more than measures in the urban population. Among age groups, those nearest the two extremes of the span of working life are most susceptible to such influences. International comparisons of the prevalence of child labour based on census statistics are especially dubious. On the whole, it is the statistics of little-developed countries that must be treated with the most cautious reserve. In such countries, where much of the work of income production is done in family enterprises, the distinction between breadwinners and others is generally less clear-cut than it is in industrialised countries, where most of the breadwinners work for wages outside the domain of the family.

The conditions that prevail in agriculture in many countries, at both low and high levels of economic development, are conducive to extreme variations in enumerations of the female labour force. It is usually the custom that women in farming families take some part, large or small, year-round or seasonal, in the production of field crops or in other farm work that is closer to the domestic sphere (such as milking, feeding livestock and kitchen gardening), in addition to keeping house, preparing meals and caring for children. The number of women counted as agricultural workers in such circumstances may be large or small, depending not only on how the census questions are worded, what is said in the instructions and who is the respondent, but also on how the roles of women are regarded in the society and what values are placed on their contributions to agricultural production on the one hand and their domestic and maternal functions on the other. In some censuses, set rules have been adopted for classifying female members of farmers' families who do not report a non-agricultural gainful occupation: either considering them all to be employed in agriculture, or counting them all as housewives not in the labour force. Such rules may be viewed as reflecting evaluations of women's roles and status in the societies concerned.

Where very few of the agricultural workers recorded in the census are women, there is a strong presumption that women's share in the agricultural labour force is understated by the statistics. In some countries, this is confirmed by information given in the census reports or by data from other sources. On the other hand, where many females are reported as active in agriculture, the statistics may give an exaggerated impression of the importance of their contribution to the labour supply, since many of them may be involved only to a small extent in agricultural work. In predominantly agricultural countries, the census measures of overall dimensions of the labour force are greatly affected by the extent to which women's participation in agriculture is reported.

Not only comparisons between countries but also measures of changes in labour force dimensions from census to census within the same country may be falsified by the factors of non-comparability, error and bias. In many countries where the concept of usual activity was the basis of labour force enumerations in early postwar censuses, the basis has been shifted in more recent censuses to activities during a specified time period. In other respects, as well, there has been a widespread tendency to sharpen the questions and definitions with more exact specifications of criteria for reporting borderline groups. It is impossible to lay down any general rule with regard to effects of such changes upon the labour force measures. The labour force enumerations in the more recent censuses were probably inflated in some countries and deflated in others, in comparison with measures provided by earlier censuses. In any country, the size of the effect, if not its direction, would be expected to vary in different age – sex groups, occupation, industry and status categories, and in the rural and urban sectors. The uncertainty of trend measures is greatest where data of only two censuses are available. Series of data from three or more censuses showing a consistent pattern of trends merit more confidence.

Adjustments can be made in the data of some censuses to improve comparability between countries or in time series for the same country. For example, where the labour force tabulations excluded groups such as armed forces, unemployed workers or persons engaged only secondarily in economic activities, and the numbers of these groups were given or a satisfactory basis for estimating them could be found, the data can be adjusted to include them in the labour force. Likewise, where groups such as women engaged only in domestic duties were included in the labour force, these can be deducted if a satisfactory basis was available. But the feasible adjustments that may be made may be far from adequate to put the data on a firm footing of comparability.

In view of what has been said, the reader may wonder how much meaning the results of a comparative study of these statistics could have. Actually, in spite of their faults, the data exhibit patterns of variations that are certainly significant and that afford valuable insights into economic and other factors affecting the levels and trends of participation in the labour force. But it is a matter of the first importance, interpreting these variations, to bear in mind the limitations of reliability and comparability of the data as well as the defects of regional coverage. The risk of being deceived by distorted reflections of the true patterns of variations will be a constant preoccupation in our analysis.

In addition to the factors mentioned above, the comparability of the measures is impaired by differences in age limits of the labour force enumerations, but this difficulty has been largely overcome by adjustments of the statistics. As a rule, no upper age limit is imposed, but there is commonly a lower age limit, and this varies: as low as 6 years or as high as 15 years among the censuses included in the present compilation. The understatement of size of the labour force due to the age limit is trivial in most cases, because few children below the specified age would be much involved in income-producing work, but the understatement may be appreciable in some cases, where the age limit does not correspond closely to the prevailing conditions of child labour. The statistics can be adjusted to a standard age reference of 10 years and over for the labour force, with estimates as necessary where a different age limit, or no limit, is imposed in the census enumerations. Errors in the estimates made for this purpose would seldom have appreciable effects upon the labour force totals for 10 years and over.

Finally, comparative analysis of rates of participation in the labour force according to age groups in different countries is hampered by differences in the age classifications found in the census tabulations (Durand, 1975).[3]

AVAILABILITY, RELIABILITY AND INTERPRETATION OF THE DATA AND OTHER RELATED PROBLEMS

The difficulties in obtaining accurate data on employment and, particularly, on unemployment and underemployment in Africa are enormous. The basic reasons for the almost complete lack of statistics on the non-regularly employed population in Africa are lack of clear definitions, lack of resources to collect these comparatively advanced statistics and lack of interest in the issue. Government authorities and industrial enterprises have no interest in a good statistical picture of the unemployment situation, as long as there is a relative abundance of manpower (unskilled), which is the case in most African countries. If the situation had been reversed, if there had been a shortage of labour and a relatively high supply of labour opportunities, the statistics would probably also have been better (Bondestam, 1973).

Availability and reliability of these statistics in Africa are well summarised in a recent Economic Commission for Africa (ECA) paper (1971), as follows:

> Actually, not enough attention has been given in the past to collecting adequate and reliable information on the economic activity of the population of Africa. Statistics related to this present special problems, which arise from the fact that the methods and concepts commonly used elsewhere are difficult to apply in an economy where the non-monetary sector is still very important and the division of labour is much less advanced. As the conditions are now, it is often difficult to ascertain whether women are economically active, nor is it easy to apply the same minimum age for employment in most African censuses as obtains elsewhere. The age limit used in some African censuses has tended to be lower than usual; 6 years of age in Libya and Algeria, 10 years in Ethiopia, and so on.

Most African countries have reasonably good records of permanent employment in the 'modern' sector: that is, urban establishments including industries, large scale farms, other large enterprises in the rural areas and the public sector. Kenya, for example, has fairly detailed statistics on number of employees by sector, industry, province, town, race, nationality, sex, occupation and earnings (see, for example, Kenya, 1971a, tables 200–218; and Kenya, 1972a). But to find statistics on unemployment one has to dig in non-government statistics, and it is difficult to find good estimates even here (Ghai, 1970).

The main reason for the lack of statistics on underemployment is the difficulty to measure the time spent by farmers and livestock-breeders on active work. Those limited statistics that exist therefore originate from fairly small sample surveys among a limited number of people. A small study in Nigeria, for example, showed that male farmers in the study area worked only 5·3 hours per day, on average (Janelid, 1971); in a region in Gabon the men spend few, if any, days in agricultural work, according to one researcher (ECA, 1971); in a Tanzanian report it has been stated that:

> Data on unemployment and underemployment are scarce. Agrarian economies typically are characterized by high degree of underemployment during portions of the year since the work force is geared to the peak period of demand in the annual round of agricultural activities. A figure for 1965 puts the rate of underemployment among wage and salary workers at 14 per cent. (Kanter, 1971)

An important subgroup of the underemployed is the irregularly employed category. This usually big group, which is mainly found in towns, is often subjectively classified as either 'employed' or 'unemployed', according to the classifier's positive or negative attitude towards the economic policy of the country.

If the concepts of employment and unemployment are vague in the context of Africa, the statistical evidence, even if it were prolific, could not be precise. For example, the various definitions of the female active population pose serious difficulties in the interpretation of statistics on this issue. Differences in the definition of female activity affect the crude economic-activity rates: that is, the number of economically active people divided by the total population. Bondestam (1973) has compared two groups of economic activity rates: the first group of countries is chosen among those with high female rates, and the second group among those with unrealistically low female rates. Consequently, the rates for both sexes are all above 50 per cent in the first group, but 30 per cent or below in the second. These differences in the female rates, he has noted, are too large to be explained by social and cultural factors only, but are rather due to the different criteria used to define and calculate the female labour force. So, for example, some countries do include females who work as unpaid persons in household enterprises; some do not. Even more important is the definition used for women in agriculture, which engages the bulge of the population. The Ethiopian example of not including women as economically active is given (see, for example, Ethiopia, 1971):

> In the age group 10–59 years, 93% of the male, 7% of the female and 50% of the total population were recorded as economically active. The two latter figures are misleading as they do not include 'housewives'. If 'housewives' were defined as an economically active category, the female economically active population would be almost the same as the male one.

In the Nigerian study (Janelid, 1971) it was found that women worked 6·4 hours per day compared to only 5·3 hours' work for men. In the West Lake region of Tanzania the men work 1,800 hours annually and women 2,600 hours in agriculture (Tanzania, 1971). Several more examples of the high female economic activity in African rural areas could be given (Boserup, 1970). Employment statistics, which exclude the working part of the female population, are therefore misleading.

Results from censuses or surveys carried out in a number of developing countries, including a few in Africa, to examine the general level of activity during both slack and busy seasons confirm the unreliability of employment surveys. A recent report observes that:

> Perhaps the most striking aspect of the survey (or for that matter census) results is that on the whole they do not conform to a widely held opinion about unemployment and under-employment in rural areas. They provide, in particular, little or no evidence for the commonly expressed view that labour has little or nothing to do for four to six months of the year in the rural sector. In Korea for example (the results show) a 40 per cent utilisation rate in the slackest period in farm households, and this rate undoubtedly exaggerates the amount of unused labour because many additional workers who would not be available all the year round are drawn in at the peak season (June) used here as the 'full utilisation' standard.
>
> The interpretation of this finding is however much more difficult. There are scattered but in total fairly numerous studies at the sub-regional or village level which suggest that the seasonal pattern in actual cultivation requirements for labour is extremely

> marked. Quite a few of these studies also suggest a low over-all rate of working, but these relate mostly to Africa. Perhaps the most important point to emphasise is that we still need a great deal more detailed and extended field work before being able to derive well-founded conclusions.
>
> One problem deserving particular mention is the treatment of activity not directly related to on-farm cultivation – which includes things like marketing, food preparation (basic food processing is often done in the home), maintenance of buildings, irrigation channels, etc. Studies show both that these activities use up quite a lot of time and that they tend to offset the seasonal pattern of on-farm cultivation work. (Turnham, 1970)

One further example will suffice to illustrate the problems of interpretation of employment and unemployment data. Killick (1978), writing on the employment and unemployment situation, generally agreed with Turnham's remarks when he commented on the reliability of employment data from Ghana, as follows:

> Finally, the state was expanding rapidly as an employer in the first half of the sixties – recorded public sector employment went up by 93,000 or 63 per cent between 1960 and 1965 and by the end of the decade stood at about the 1965 level.[4] This increase alone was equivalent to well over half the unemployment recorded in 1960.
>
> While they are the best sources available, not too much weight should be placed on inter-censal comparisons. That the 1960 and 1970 censuses were held at the same time of year and the relevant questions were the same reduces the difficulties, as does the absence of obvious abnormalities in economic conditions at the time of both censuses. However, there are difficult conceptual and practical problems in measuring unemployment in countries like Ghana, arising from imperfect specialization and the existence of various informal types of work. One may be particularly uneasy about the major changes recorded in female participation and unemployment between the two censal years, which strongly influenced the results reported here.
>
> One obvious limitation of census data is that they can tell nothing about trends within the decade. Given that the expansion of public sector employment was concentrated in the first half of the decade, it is probable that unemployment went down in that period and then rose in the later sixties. This is consistent with evidence that secondary school leavers began to experience growing difficulties in finding work towards the end of the decade. It may well be, then, that governments had a more accurate impression of what was going on than the census figures suggest, although

alarmist views of the size of the problem are contradicted by evidence that there were labour shortages in the rural economy and that, contrary to official statements, there was no unemployment among university graduates at the beginning of the seventies.

To suggest that unemployment may not be the problem it is often thought to be is not, of course, to deny that it is a problem at all. On the contrary, to have nearly 15 per cent of the non-agricultural labour force unemployed is both a waste of productive resources and a symptom of real social distress. What is more, unemployment, as conventionally defined, records only part of the story – it leaves out the disguised unemployment and underemployment that are also prevalent. Even though there is little hard data, it is probable that much seasonal unemployment occurs in agriculture during the dry months of the year, and much of the petty trading in the towns can best be regarded as disguised unemployment.

The statistical evidence concerning unemployment derives mostly from registered unemployment and, for that reason, refers mainly to the unemployment situation in urban areas. Bondestam (1972a), commenting on the reliability of unemployment statistics and their lack of representativeness in Ethiopia, has remarked that the statistics are undoubtedly reliable but the information only covers the voluntary registration of job-seekers, the vacancies willingly notified by employers and the placements effected in the three months' period. Since both the registration and vacancy notifications are voluntary, the information does not claim to cover the national employment and unemployment situation.

The statistical analysis of unemployment data from several sources of registered unemployment reveals that some countries record for their urban areas rather low unemployment rates (Turnham, 1970; ILO, *Yearbook of Labour Statistics*, annual; also Mouly and Costa, 1974). This does not necessarily imply a close approach to full employment; rather, it more likely reflects the fact that registration has not yet become customary. It is impossible to guess by how much these various percentages should be increased to show the true situation; in all developing countries registered unemployment represents no more than the above-water tip of the iceberg of unemployment and underemployment; but just as icebergs differ in their configuration, so does the difference between recorded unemployment and unrecorded underutilisation of labour vary from country to country. We only know that the difference is great.

The statistics also reveal almost everywhere a much heavier incidence of unemployment among young people in the 15–24 age group. In several countries, the rate in this age group is more than double the national average. This is accounted for partly by the heavy preponderance of young people in the migrants from countryside to towns, and partly by the fact that the young are more inclined to register for work than the middle-aged, who may have lost interest or hope in job-seeking. In some countries the ranks of the unemployed contain a large component of persons who have passed through secondary school and are reluctant to accept anything less than relatively well-paid posts. Educated persons also tend to be concentrated in urban areas, where there are fewer opportunities, than in farm zones for employment as unpaid family workers. Nonetheless, although the contrast may be exaggerated by these statistics, the concentration of the most severe unemployment among the young people is a matter of common observation.

As a result of these deficiencies in measuring unemployment, one finds that national employment figures or statistics in African countries, while indicating the same level of unemployment rates, may to some extent give a distorted picture of the unemployment situation. This is because the figures are defined and collected differently; one figure may tend to exaggerate true unemployment, whereas the other may tend to underestimate it. The most important differences come in the way the figures are collected. Figures based on registered unemployment naturally tend to be lower than those based on sample surveys. Registered unemployment includes only those who job-hunt through the state employment exchanges or who need to register to qualify for social benefits.

Mouly and Costa (1974) have further noted that, for a few countries, sample surveys of unemployment have been conducted regularly over a period of years (published in the ILO's *Yearbook of Labour Statistics*). It is difficult to draw any firm conclusions from these time series. In some of the countries the unemployment situation (expressed as a percentage of the labour force) appears to have deteriorated; in others it has improved, while in yet others it has shown little change. Some of the apparent deterioration (or improvement) may be due to administrative changes during the period. In any case, the comparatively small number of countries providing these data cannot be regarded as necessarily representative of developing countries as a whole.

Finally, as with every census, there are general conceptual problems and non-sampling errors that impair the usefulness of the statistical results. Dey and Mogil (1973) have agreed and commented on the Tanzanian situation that:

> In relating the census data to its major uses, it is apparent that the planning value of the data is very limited. The inadequate definitions of 'employment' and especially 'unemployment' do not

permit conclusions to be reached about the nature and extent of 'unemployment' in Tanzania. Equally, manpower planners' use of the data is limited, less aggregated occupational classification is required for manpower projections and national accounting; in addition, misclassification will further impair the usefulness of the data.

This brief review of the statistical evidence on unemployment shows how fragmentary the material is and how little claim it can make to being representative of the true situation. All that one can say with any certainty is that the statistics understate the magnitude of the problem. In many countries they probably do not even cover the majority of those who are wholly unemployed and are seeking work; they make no claim to measure the volume of visible and invisible underemployment. Nor do they throw any light on the question of whether unemployment and underemployment are becoming worse. Again, one can say with confidence that, whatever the real trend in the magnitudes may be, public opinion is becoming more deeply concerned with the existence of this problem.

USES OF LABOUR STATISTICS

The uses of the statistics of employment, unemployment and the labour force are enormous:

(1) They provide an adequate statistical basis for the analysis of economic and social problems of employment and unemployment, and in particular for the formulation and application of policies designed to maintain full employment and to promote economic development. However, for an employment policy it is important that data on incomes or compensation should correspond as far as possible to data on employment. As a rough approximation, income from each activity, or the income equivalents of different tasks, should be matched by corresponding labour inputs (in terms of man-hours) in order to assess the relative importance of different activities and their capacity to generate income.

(2) The Feldman – Mahalanobis model of planning clearly specifies employment targets along with the targets of different products to be achieved. Thus, once the basic statistics of employment and the labour force are available, one can estimate the future requirements of labour force commensurate with the planned growth of output in various industries.

(3) The reliable statistics of employment, unemployment and the labour force are crucial for estimating labour productivity in

different industries as well as that of the nation. Productivity analysis is the basis of planning for economic development and also of international comparisons.

(4) The reliable statistics of employment, unemployment and labour give an idea of the available resources in the country and form the basis of any policy towards optimisation. The country can choose a labour-intensive technique if labour is abundant, or carry out a labour-saving programme if capital is abundant and labour is scarce. Such statistics will also reveal the economic ills in the country. If unemployment is rising, it will lead to many socioeconomic problems and will call for urgent remedial measures.

(5) When the statistics of employment are available in suitable form, one can determine the purchasing power and the consumption patterns. When unemployment is rising, the purchasing power of the people at large falls, and vice versa. The consumption pattern also shifts from luxury and comforts to mere necessities of life, provided that the distribution of income remains the same as before.

(6) Unemployment data obtained from the unemployment register can be used in the construction of life tables with 'remaining on the register' in the place of survival, and 'departure from the register' in the place of death. A feature of this kind of table is that the expectation of remaining on the register rises with the time already spent on it. This phenomenon shows the depressing effect on the individual of a protracted spell of unemployment.

(7) The statistics on employment, unemployment and education can be used for manpower-planning. We should emphasise that employment difficulties are not caused by population growth alone, but also by increases in the labour supply relative to labour demand. Expansion in labour demand depends upon a variety of economic factors, including aggregate output growth, sectoral shifts in the composition of output, changes in the relative price of labour, changes in the technical capital – output and capital – labour ratios, and a variety of institutional characteristics, including government wage levels and minimum wage legislation. Changes in the labour supply are also affected by these factors, as well as by demographic variables.

These increasing difficulties bring into focus manpower-planning as a logical area of interest and the need for manpower statistics as a tool for planning. The goal of manpower-planning is to project labour force supplies and demands and to make recommendations for government policy to help reconcile the two.

CONCLUDING REMARKS

Employment statistics as a measure of the degree of achievement of the economic goals of a nation in Africa can be misleading. Underemployed farmers can often be better off than full-time engaged wage-earners, quasi-employed street vendors and prostitutes, who often make more money than underpaid servants, who may work up to 10 – 15 hours a day. Statistics on employment can therefore reflect only the level of engagement of a population employed in production and services. For employment statistics to become meaningful, they have to be accompanied by income statistics and their distribution. This will be discussed at length later.

The past few years, however, have undoubtedly seen great improvements in the quality of employment and unemployment statistics. Examination of the series for numerous countries discloses the development of more dependable sources, more complete reporting, greater refinements of statistical methods, more detail in reported data, and greater use of benchmark data in sample selection and the correction of trend bias. This improvement has, however, been patchy. Many series still in use are crudely constructed and of doubtful value as a basis for policy formulation.

Finally, despite the recommendations of international agencies, national series of employment and unemployment are far from comparable, even in the most advanced countries of the world, not to speak of African countries that are at the preliminary stage of data collection. They differ in the definition and coverage of the employed and the unemployed.

NOTES: CHAPTER 7

1 See, for example, Singer (1975), who has compared Kenya and Great Britain. He has noted that, because its population is growing so much faster, Kenya must find, per million of population, eight times as many additional productive-employment opportunities as Britain in order to keep unemployment from rising. It must do this with roughly one-fifteenth of the investment resources. Fundamentally, what is required is a lower rate of population increase or much more labour-intensive methods of production or, preferably, both.

2 In 1954, the International Labour Organisation (ILO) defined unemployment to include (2) and (3) but not (1). See, for example, a discussion on problems of measurement in *The Economist* (1978, pp. 90 – 1).

3 For a full discussion of this section, see, for example, Durand (1975).

4 Their limitations are clear from the fact that the official 1970 figures of recorded employment represented only 12 per cent of total employment as recorded in the 1970 census (after adjustment), and only 28 per cent of non-agricultural employment. Moreover, the trends shown are also likely to be unreliable because of incomplete and variable reporting coverage. These statistics show total recorded employment to have increased 17 per cent in 1960 – 70, whereas the census returns

indicate a 41 per cent increase in non-agricultural employment in the same period. See Williams and Ntim (1973) and Killick (1966, ch. 6; and 1972) for discussion of the reliability and interpretation of the labour statistics.

REFERENCES AND FURTHER READING

Addo, N. O. (1967), 'Demographic aspects of manpower and employment in Ghana', *Economic Bulletin of Ghana*, no. 3.

Addo, N. O. (1972), 'Employment and labour supply on Ghana's cocoa farms in the pre and post Aliens Compliance Order era', *Economic Bulletin of Ghana*, no. 4.

Amsden, A. H. (1971), *International Firms and Labour in Kenya, 1945–70* (London: Frank Cass).

Anker, R. and Knowles, J. C. (1977), 'Female labour force participation in Kenya', International Labour Organisation, Geneva (mimeo.).

Annable, J. E. Jr (1972), 'Internal migration and urban unemployment in low-income countries: a problem in simultaneous equations', *Oxford Economic Papers*, new series, pp. 399–412.

Bienefeld, M. A. (1975), 'Employment and unemployment', in B. Dasgupta and D. Seers (eds), *Statistical Policy in Less Developed Countries*, IDS No. 114 (Brighton: Institute of Development Studies, University of Sussex).

Bienefeld, M. A. and Godfrey, E. M. (1975), 'Statistical problems of measuring unemployment and the informal sector', paper presented at the conference on Statistical Policy in Less Developed Countries, Institute of Development Studies, University of Sussex, Brighton (12–16 May), IDS No. 114.

Blaug, M. (1974), 'Employment and unemployment in Ethiopia', *International Labour Review*, vol. 110, no. 2, pp. 117–44.

Bondestam, L. (1972a), *Reliability of Population Data in Ethiopia* (Addis Ababa: Central Statistics Office).

Bondestam, L. (1972b), *Unemployment in Addis Ababa, Asmara and Dire Dawa* (Addis Ababa: Central Statistics Office).

Bondestam, L. (1972c), 'Underdevelopment and economic growth in Ethiopia', paper prepared for the seminar on Population and Economic Growth in Africa, Leiden.

Bondestam, L. (1973), *Some Notes on African Statistics: Collection, Reliability and Interpretation*, Research Report No. 18 (Uppsala: Scandinavian Institute of African Studies).

Boserup, E. (1965), *The Conditions of Agricultural Growth: The Economics of Agrarian Change under Population Pressure* (Chicago, Ill.: Aldine Publishing).

Boserup, E. (1970), *Women's Role in Economic Development* (London: Allen & Unwin).

Boserup, E. (1972), 'Women: the neglected human resources for African development', article prepared for the special issue on Women in Development of the *Canadian Journal of African Studies* by the Human Resources Development Division, Economic Commission for Africa, Addis Ababa.

Brown, R. (1965), *Report of the Government of Sierra Leone on the Manpower Situation* (London: Ministry of Overseas Development). (In Fourah Bay College Library, Freetown.)

Caldwell, J. C. (1977), 'The economic rationality of high fertility: an investigation illustrated with Nigerian survey data', *Population Studies*, vol. 31, no. 1, pp. 5 – 27.

Centre for Developing Area Studies (undated), *Bibliography on Labour Force Participation Rates in Nigeria, 1970 – 1976* (Montreal: McGill University).

Child, F. C. (1973a), 'Employment, technology and growth – role of the intermediate sector', paper prepared for the conference on Strategies for Increasing Productive Employment in African Countries, Limuru.

Child, F. C. (1973b), *An Empirical Study of Small-Scale Rural Industry in Kenya*, Working Paper No. 127 (Nairobi: Institute for Development Studies, University of Nairobi).

Child, F. C. (1973c), *Entrepreneurship, Management and Labour in a Society in Transition*, Working Paper No. 128 (Nairobi: Institute for Development Studies, University of Nairobi).

Child, F. C. (1973d), *Programs and Policies for Promoting Growth of the Intermediate Sector in Kenya*, Working Paper No. 130 (Nairobi: Institute for Development Studies, University of Nairobi).

Cox-George, N. A. (1950), 'An essay on employment and unemployment in Sierra Leone in 1948', unpublished paper. (In Fourah Bay College Library, Freetown.)

Dawson, J. L. M. (1963), 'Traditional values and work efficiency in a West African mine labour force', *Occupational Psychology*, 37, pp. 209 – 18.

de Graft-Johnson, K. T. (1970), 'Ghana: population growth and its relationship to employment and unemployment in Ghana', in *The Demographic Transition in Tropical Africa*, proceedings of an Expert Group Meeting, No. 142 (Paris: Organisation for Economic Co-operation and Development).

de Graft-Johnson, K. T. (1975), *Factors Affecting Labour Force Participation Rates in Ghana, 1970* (Geneva: International Labour Organisation).

Dewney, J. C. (1967a), 'The distribution of employment in Sierra Leone', *Sierra Leone Geographical Journal*, no. 11, pp. 51 – 9.

Dewney, J. C. (1967b), 'Employment and unemployment in Sierra Leone', *Economic Trends*, new series, no. 1, pp. 1 – 8.

Dey, A. K. and Mogil, N. (1973), 'Economic characteristics', in Egero and Henin (1973).

Diejomaoh, V. P. (undated), *An Overview of the Employment Problem in West Africa*, Occasional Paper (London: Commonwealth Secretariat Youth Division).

Doctor, K. C. and Gallis, H. (1966), 'Size and characteristics of wage employment in Africa', *International Labour Review*, vol. 93, no. 2.

Durand, J. D. (1975), *The Labour Force in Economic Development: A Comparison of International Census Data, 1946 – 1966* (Princeton, NJ: Princeton University Press).

Economic Commission for Africa (ECA) (1968), *Demographic Handbook for Africa* (Addis Ababa).

ECA (1971), *Rapport de la mission interdisciplinaire sur le développement*

rural intègre de la région de la N'Gouine, Gabon, Document No. M71-1185 (Addis Ababa).
Edgren, G. (1976), 'On the politics and mechanics of wage control: a comparative study of experience in some Western and African countries', paper presented at the East African Countries Research Symposium on Industrial Relations and Employment, Institute for Labour Studies, Lusaka.
Egero, B. (1973), 'Non-sampling errors', in Egero and Henin (1973).
Egero, B. and Henin, R. A. (1973), *The Population of Tanzania: An Analysis of the 1967 Census*, Census Vol. 6 (Dar es Salaam: Bureau of Statistics).
El Shafei, A. M. N. (1960), 'The current labour force sample survey in Egypt', *International Labour Review*, vol. 82, no. 5, pp. 432–49.
Ethiopia (1971), 'Population of Ethiopia: results from the national sample survey, 1st round, 1964–67', *Statistical Bulletin*, no. 6 (November) (Addis Ababa: Central Statistics Office).
Ethiopia (1972), *Labour Exchange Information* (Addis Ababa: Central Statistics Office, Ministry of National Community Development, quarterly).
Fajana, O. (1975), 'Intraindustry wage differentials in Nigeria', *Journal of Developing Areas*, 10 (July).
Food and Agriculture Organisation (FAO) (1977), *Monthly Bulletin of Agriculture, Economics and Statistics*, vol. 25 (Rome).
FAO (various dates), *Report on Employment and Unemployment* (Rome).
Fowler, R. F. (1968), *Duration of Unemployment on the Register of Wholly Unemployed*, Studies in Official Statistics Research Series No. 1 (London: HMSO).
Frankman, M. and Charlie, E. (1973), 'Employment in the service sector in the sub-Saharan', *Journal of Modern African Studies*, vol. 11, no. 2.
Friedmann, J. and Sullivan, F. (1974), 'The absorption of labor in the urban economy: the case of developing countries', *Economic Development and Cultural Change*, pp. 385–413.
Ghai, D. P. (1967), *Analytical Aspects of an Incomes Policy for Kenya*, Discussion Paper No. 50 (Nairobi: Institute for Development Studies, University of Nairobi).
Ghai, D. P. (1968), *Strategy for a Public Sector Wage Policy in Kenya*, Discussion Paper No. 65 (Nairobi: Institute for Development Studies, University of Nairobi).
Ghai, D. P. (1970), *Employment, Performance Prospects and Policies in Kenya*, Discussion Paper No. 99 (Nairobi: Institute for Development Studies, University of Nairobi).
Ghai, D. P. (1973), 'Incomes policy in Kenya: need, criteria and machinery', in Jolly *et al.* (1973), pp. 153–69.
Ghana (1961), *Survey of High Level Manpower in Ghana, 1960* (Accra: Manpower Unit, Ministry of Economic Planning).
Ghana (1971), *High Level and Skilled Manpower Survey in Ghana, 1968, and Assessment of Manpower Situation* (Accra: Development Planning Secretariat).
Ghana (annual), *Labour Statistics* (Accra: Central Bureau of Statistics).
Ghana (monthly), 'Employment market situation report', Labour

Department, Accra (mimeo.).
Greenstreet, M. (1971), 'Employment of women in Ghana', *International Labour Review*, vol. 103, no. 2, pp. 117–30.
Gronau, R. (1973a), 'The effect of children on the housewife's value of time', *Journal of Political Economy*, vol. 81, no. 2, pt 2, pp. 168–99.
Gronau, R. (1973b), 'The intra-family allocation of time: the value of housewife's time', *American Economic Review*, vol. 63, pp. 634–51.
Gunnor, M. (1968), *Asian Drama* (New York: Pantheon).
Gutkind, P. C. W. (1967), 'The energy of despair: social organisation of the unemployed in two African cities: Lagos and Nairobi', *Civilizations*, 17.
Gutkind, P. C. W. (1972), *Unemployment in Africa* (Montreal: Centre for Developing Area Studies, McGill University).
Gutkind, P. C. W. (1973), 'The unemployed and poor in urban Africa', in Jolly *et al.* (1973).
Gutkind, P. C. W. (1977), *Bibliography on Unemployment with Special Reference to Africa* (Montreal: Centre for Developing Area Studies, McGill University).
Gwyer, G. (1972), *Trends in Kenya Agriculture in Relation to Employment*, Discussion Paper No. 153 (Nairobi: Institute for Development Studies, University of Nairobi).
Harris, J. R. (1971), 'Wage rate determination with limited supplies of labour in developing countries', *Journal of Development Studies*, 7.
Harris, J. R. and Todaro, M. P. (1969), 'Wages, industrial employment and labour productivity', *Eastern Africa Economic Review*, 1.
Hart, J. K. (1973), 'Informal income opportunities and urban employment in Ghana', *Journal of Modern African Studies*, 11.
Henning, P. H. and House, W. J. (1975), 'The problems of slow employment growth in less developed economies: some observations from the Kenyan experience', *Journal of Eastern Africa Research and Development*, 2.
House, W. J. (1973a), 'Market structure and industry performance: the case of Kenya', *Oxford Economic Papers*, new series, 25, pp. 405–19.
House, W. J. (1973b), 'Wages, employment and productivity in Kenya: some further evidence', *Eastern Africa Economic Review*, 5.
House, W. J. and Rempel, H. (1975), *An Analysis of the Variation in Modern Sector Earnings Among the Districts and Major Urban Centres in Kenya*, Working Paper No. 243 (Nairobi: Institute for Development Studies, University of Nairobi).
House, W. J. and Rempel, H. (1976a), 'The determinants of and changes in the structure of wages and employment in the manufacturing sector of the Kenya economy, 1967–1972', *Journal of Development Economics*, 3, pp. 83–98.
House, W. J. and Rempel, H. (1976b), 'The impact of unionization on negotiated wages in the manufacturing sector in Kenya', *Oxford Bulletin of Economics and Statistics*, 38, pp. 111–24.
House, W. J. and Rempel, H. (1976c), 'Labour market segmentation in Kenya', *Eastern Africa Economic Review*, 8.
House, W. J. and Rempel, H. (1977), 'Labour market pressures and wage determination in less developed countries: the case of Kenya', *Economic Development and Cultural Change*, 25.

Humphrey, D. (1940), 'Alleged "additional workers" in the measurement of unemployment', *Journal of Political Economy*, vol. 48, no. 3, pp. 412–19.
Hunter, G. (1972), 'Employment policy in tropical Africa: the need for radical revision', *International Labour Review*, vol. 105, no. 1, pp. 35–57.
Hunter, J. B. K. (1969), 'The development of the labour market in Kenya', in I. G. Stewart (ed.), *Economic Development and Structural Change* (Edinburgh: Edinburgh University Press), pp. 113–37.
International Bank for Reconstruction and Development (1963), *The Economic Development of Kenya* (Baltimore, Md: Johns Hopkins).
International Labour Organisation (ILO) (1966a), 'General report on labour statistics', Report 1, Eleventh International Conference of Labour Statisticians, Geneva.
ILO (1966b), 'Measurement of underemployment – concepts and methods', Report 4, Eleventh International Conference of Labour Statisticians, Geneva.
ILO (1966c), *Application of Sampling in Labour Statistics* (Geneva).
ILO (1966d), *International Labour Review Technical Guide* (Geneva), section 2, 'Unemployment'; section 3, 'Employment'.
ILO (1970a), *Labour Force Projection, 1965–1985. Part 2: Africa* (Geneva).
ILO (1970b), 'Manpower assessment and the planning situation in Sierra Leone', report of ILO Regional Manpower Adviser presented to the Government of Sierra Leone, Freetown.
ILO (1972), *Employment, Incomes and Equality: A Strategy for Increasing Productive Employment in Kenya* (Geneva).
ILO (1973a), *Employment and Unemployment in Ethiopia* (Geneva).
ILO (1973b), 'Technical memorandum to the government of Sierra Leone on the re-organisation of the Employment Service', report presented to the government of Sierra Leone, Freetown.
ILO (1973c), 'Report and recommendations on the Central Employment Exchange, Freetown', report presented to the government of Sierra Leone, Freetown.
ILO (1976), *International Recommendations on Labour Statistics* (Geneva).
ILO (annual), *Yearbook of Labour Statistics* (Geneva).
ILO (monthly), *Statistical Supplement to the International Labour Review* (Geneva).
ILO (quarterly), *Bulletin of Labour Statistics*, with supplements in the intervening months (Geneva). (This bulletin complements the annual data given in the yearbook.)
Inukai, I. (1973), 'The legal framework for the informal sector', paper prepared for the conference on Strategies for Increasing Productive Employment in African Countries, Limuru.
Ivory Coast (1967), *Employment* (Abidjan: Ministry of Planning).
Janelid, I. (1971), 'The socio-economic role of farm women in agricultural and national development in West Africa', paper prepared for FAO conference on Agricultural Education at the Intermediate Level, Zaria, table 1.
Johnson, G. E. and Whitelaw, W. E. (1974), 'Urban–rural income transfers in Kenya: an estimated-remittances function', *Economic Development and*

Cultural Change, 22, pp. 473–9.
Johnson, H. M. A. (1941), 'Labour conditions in Sierra Leone', *International Labour Report*, 44.
Jolly, R., DeKadt, E., Singer, H. and Wilson, F. (eds) (1973), *Third World Employment: Problems and Strategy* (Harmondsworth: Penguin).
Jolly, R., Seers, D. and Singer, H. (1974), *The Pilot Employment Missions and the Lessons of the Kenya Mission*, IDS No. 111R (Brighton: Institute of Development Studies, University of Sussex).
Kanter, J. (1971), 'Tanzania: a case study', paper prepared for the African Population Conference, Accra.
Kenya (1969), *Development Plan, 1970–1974* (Nairobi: Government Printer).
Kenya (1971a), *Statistical Abstract* (Nairobi: Statistics Division, Ministry of Finance and Economic Planning).
Kenya (1971b), *Report of the Commission of Inquiry: Public Service Structure and Renumeration Commission, 1970–71*, chaired by D. N. Ndegwa (Nairobi: Government Printer).
Kenya (1971c), *Register of Manufacturing Firms, 1970* (Nairobi: Statistics Division, Ministry of Finance and Economic Planning/Government Printer).
Kenya (1971d), *Employment and Earnings, 1963–1967* (Nairobi: Statistics Division, Ministry of Finance and Economic Planning).
Kenya (1972a), *Economic Survey* (Nairobi: Statistics Division, Ministry of Finance and Economic Planning, June).
Kenya (1972b), *Census of Industrial Production, 1967* (Nairobi: Statistics Division, Ministry of Finance and Economic Planning).
Kenya (1972c), *Employment and Earnings in the Modern Sector, 1968–1970* (Nairobi: Statistics Division, Ministry of Finance and Economic Planning).
Kenya (1973a), *Employment and Earnings in the Modern Sector, 1971* (Nairobi: Statistics Division, Ministry of Finance and Economic Planning).
Kenya (1973b), *Sessional Paper on Employment*, no. 10.
Kenya (1974a), *Development Plan, 1974–1978* (Nairobi: Government Printer).
Kenya (1974b), *Statistical Abstract, 1973* (Nairobi: Statistics Division, Ministry of Finance and Economic Planning/Government Printer).
Killick, T. (1966), ch. 6 in W. Birmingham *et al.* (eds), *A Study of Contemporary Ghana*, Vol. I (London: Allen & Unwin).
Killick, T. (1972), 'Unemployment as an indicator of prosperity: a statistical curiosity', *Economic Bulletin of Ghana*, no. 3.
Killick, T. (1978), *Development Economics in Action: A Study of Economic Policies in Ghana* (London: Heinemann).
Kinyanjui, P. K. (1973), 'Education, training and employment of secondary school leavers in Kenya', *Manpower and Unemployment Research in Africa: A Newsletter*, 9.
Knight, J. B. (1967), 'The determination of wages and salaries in Uganda', *Bulletin of the Oxford University Institute of Economics and Statistics*, 29.
Knight, J. B. (1968), 'Earnings, employment, education and income distribution in Uganda', *Bulletin of the Oxford University Institute of Economics and Statistics*, 30.
Langdon, S. (1975), 'Multinational corporations, taste transfer and

underdevelopment: a case study from Kenya', *Review of African Political Economy*, 1.
Lawson, R. M. (1968), 'The traditional utilization of labour in agriculture on the lower Volta, Ghana', *Economic Bulletin of Ghana*, vol. 12, no. 1, pp. 55–6.
Leys, C. (1975), *Underdevelopment in Kenya: The Political Economy of Neo-colonialism* (London: Heinemann).
Love, J. (1973), 'Unemployment in less developed countries', staff paper presented at a seminar, Economic Commission for Africa, Addis Ababa.
Mahbub ul Hag (1971), 'Employment in the 1970s: a new perspective', *International Development Review*, no. 4.
Malawi (quarterly), *Quarterly Digest of Statistics* (Zomba: Ministry of Finance).
Merrick, T. W. (1976), 'Employment and earnings in the informal sector in Brazil: the case of Belo Horizonte', *Journal of Developing Areas*, 10, pp. 337–53.
Moses, J. E. and Bottomley, A. (1968), 'Wage rate determination with limited supplies of labour in developing countries', *Journal of Development Studies*, 4.
Mouly, J. and Costa, E. (1974), *Employment Policies in Developing Countries* (London: Allen & Unwin).
Nagi, M. N. (1971), *Labour Force and Employment in Egypt* (New York: Praeger).
Myrdal, Gunnar (1968), *Asian Drama* (New York: Pantheon).
Nigam, C. S. and Singer, H. (1974), 'Labour turnover and employment: some evidence from Kenya', *International Labour Review*, vol. 110, no. 6, pp. 479–94.
Pack, H. (1972), 'Employment and productivity in Kenyan manufacturing', *Eastern Africa Economic Review*, 4.
Rempel, H. (1970), 'Labour migration into urban centres and urban unemployment in Kenya', unpublished Ph.D. dissertation, Department of Economics, University of Wisconsin.
Rempel, H. (1974a), *An Estimate of Kenya's Labour Force*, Working Paper No. 159 (Nairobi: Institute for Development Studies, University of Nairobi).
Rempel, H. (1974b), *The Nature and Extent of Migration into Kenya's Towns*, Working Paper No. 160 (Nairobi: Institute for Development Studies, University of Nairobi).
Rempel, H. (1975), 'Rural to urban migration as a means to greater equity', paper read at a Canadian Association of African Studies meeting, York University.
Rempel, H. and House, W. (1978), *The Kenya Employment Problem: An Analysis of the Modern Sector Labour Market* (Nairobi: University Press).
Rodgers, G. B. (1975), 'Nutritionally based wage determination in the low-income labour market', *Oxford Economic Papers*, new series, 27, pp. 61–81.
Rourke, B. E. and Obeng, F. A. (1973), 'Seasonality in the employment of casual agricultural labour in Ghana', *Economic Bulletin of Ghana*, no. 3.
Rudd, K. (1972), *Programme of Manpower Information, Assessment and*

Planning (Freetown: Central Planning Unit, Ministry of Development and Economic Planning).

Rudd, K. (1973a), *Population, Manpower and Employment, 1973: View of Trends and Problems* (Freetown: Central Planning Unit, Ministry of Development and Economic Planning).

Rudd, K. (1973b), *Survey of Technical Institutes and Trade Centres, 1972* (Freetown: Central Planning Unit, Ministry of Development and Economic Planning).

Rudd, K. (1974), 'Human resources', in *National Development Plan, 1974/75 – 1978/79* (Freetown: Central Planning Unit, Ministry of Development and Economic Planning), ch. 3, pp. 19 – 28.

Rudd, K. (1976), 'The role of manpower in plan implications', lecture notes to planning officers of the Central Planning Unit, Ministry of Development and Economic Planning, Freetown (mimeo.).

Rudd, K. (1977), *Status of Human Resources Planning with Recommendations for the Remaining Period of the National Development Plan* (Freetown: Central Planning Unit, Ministry of Development and Economic Planning).

Rudd, K. and Sesay, J. B. (1973), *Survey of Vacancies in the Public Service, 1970 – 72* (Freetown: Central Planning Unit, Ministry of Development and Economic Planning).

Sandbrook, R. (1975), *Proletarians and African Capitalism: The Kenyan Case, 1960 – 1972* (Cambridge: Cambridge University Press).

Senga, W. M. (1973), 'Wages, market imperfections and labour absorption in Kenya manufacturing industries', *Eastern Africa Economic Review*, 5.

Sheehan, G. (1976), *Labour Force Participation Rates in Khartoum* (Geneva: International Labour Organisation).

Sierra Leone (undated), *Annual Report* (Freetown: Labour Department/Government Printer). (Department established 1939.)

Sierra Leone National Commission for the United Nations Educational, Scientific and Cultural Organisation (UNESCO) (1974), *Education, Training and Employment Opportunities for Women in Sierra Leone* (Freetown: Ministry of Education).

Singer, H. W. (1975), *The Strategy of International Development: Essays in the Economics of Backwardness* (New York: International Arts and Sciences Press).

Standing, G. and Sheehan, G. (1976), *Economic Activity of Women in Nigeria* (Geneva: International Labour Organisation).

Steel, W. F. and Mabey, S. J. (1973), *Unemployment and Income Distribution in Ghana since 1960: An Annotated Bibliography* (Montreal: Centre for Developing Area Studies, McGill University).

Strassmann, W. P. (1970), 'Construction, productivity and employment in developing countries', *International Labour Review*, p. 508.

Taira, K. (1966), 'Wage differentials in developing countries: a survey of findings', *International Labour Review*, 93.

Tanzania (1971), 'Country report on Tanzania', prepared for the conference of the United Nations Commission on the Status of Women, Gabon.

The Economist (1978), 'Europe's rising unemployment', schools brief (3 March), pp. 90 – 1. (Published weekly, London.)

Todaro, M. P. (1969), 'A model of labor migration and urban unemployment in less developed countries', *American Economic Review*, 59.

Turay, B. J. (1973), 'Factors influencing choice of career among boys and girls', unpublished Dip.Ed. dissertation, Fourah Bay College, University of Sierra Leone, Freetown.

Turnham, O. (1971), *The Employment Problem in Less Developed Countries* (Paris: Organisation for Economic Co-operation and Development).

Turnham, O., assisted by I. Jaegar (1970), *The Employment Problem in Less Developed Countries: A Review of Evidence* (Paris: Organisation for Economic Co-operation and Development).

United Nations (1967a), *Statistical Yearbook* (New York: UN Statistical Office).

United Nations (1967b), *Supplement to the Statistical Yearbook* (New York: UN Statistical Office).

United Nations (1967c), *Monthly Bulletin of Statistics* (New York: UN Statistical Office).

United Nations (undated), *Principles and Methods for the 1970 Population Censuses*, ST/STAT/SER.M/44 (New York: UN Statistical Office).

Wachtel, H. M. and Betsey, C. (1972), 'Employment at low wages', *Review of Economics and Statistics*, 54, pp. 121 – 9.

Webb, R. (1973), 'Income and employment in the urban traditional sector in Peru' (mimeo.).

Williams, T. D. and Ntim, S. M. (1973), 'Public employment centres as a source of data on unemployment in Ghana', *Economic Bulletin of Ghana*, no. 1.

Young, A. S. (1977), 'The employment of our graduates', University of Sierra Leone (mimeo.).

CHAPTER 8

Social Security Statistics

The major advances over the past twenty years in the collection, analysis and uses of social statistics on such topics as population and education have been much more striking than for social security statistics in Africa. While there has been a steady improvement in the collection and publication of statistics on education and population, the statistical coverage of social security has not received the same attention. There has not been a comparable increase in the range of statistics available, nor have existing data been published with the same degree of regularity. The statistics have not been brought together so conveniently, nor presented with the same regard for consistency and comprehensiveness. Nevertheless, in this chapter we shall attempt to explain the nature and scope of social security statistics and the sources, uses and limitations of the data from social security systems.

SCOPE AND NATURE OF SOCIAL SECURITY SYSTEMS

There is no universally agreed definition of the precise concept of social security or of the criteria to observe in identifying its scope. There are in fact wide differences of institutional arrangements as well as different patterns in the evolution of social programmes in various countries. These developments have conditioned national concepts as to what elements comprise social security. There is also considerable confusion in terminology; for example, terms such as social security, social welfare and social services have been used interchangeably to mean the same thing, and at other times to mean different things. As a first step in defining social security systems and their scope, therefore, we have to define the terms social security, social welfare and social services.

Social welfare and *social services* include a great variety of measures for the promotion of welfare through community effort, whereas *social security* is restricted to schemes or services set up under public legislation, conferring specific individual rights and designed to provide a subsistence income to breadwinners in case of involuntary interruption or loss of earning capacity. This general definition includes social insurance, public service and social assistance.

In defining social security, we also have to distinguish between the terms social insurance, public service and social assistance. Under *social insurance* principle, individuals are entitled to benefits on the basis of contributions to an insurance fund that have been paid by the covered persons, or by their employers, on their behalf, or both. Under the stated contingencies, the benefits must, by law, be paid to the individuals covered; possession of private means does not affect title to the benefit, and the benefits are related to the size of the contributions and the length of contribution period.

Under the *public service* concept, title benefit is conferred on the population generally at the expense of the community and without prior contributions by the covered persons (except in the form of taxes); the same benefit is available to all who qualify, and the possession of private means does not affect title to it. Under *social assistance*, benefits are provided at the expense of the community and without prior contributions by individuals who are in need. The benefits are calculated to help the individual attain a certain minimum standard of living through the application of a means test.

A study of social security in Africa (ILO, 1969; Mouton, 1975) reveals wide differences between countries in the date at which the various systems were introduced as well as in the extent of their development, the nature of the schemes or measures applied, their rate of growth and their particular tendencies.

This obvious heterogeneity should not, however, conceal certain general similarities, nor the convergence of the various systems that can be discerned and which can be expected to become more pronounced in the fairly near future.

A far greater similarity is apparent among certain clearly-defined groups of countries where one constantly encounters developments in social security systems that are comparable in terms of both time and nature. Despite the great dearth of statistics, it has, first, been possible to show that the number of contingencies concerned tends to increase more or less rapidly, but the protection extended remains confined to employees. Secondly, it has also been shown that the following types of schemes have come under social security schemes in many countries: (1) compulsory social insurance, (2) certain voluntary social-insurance schemes, (3) special schemes for public employees, (4) family allowances, (5) public assistance, (6) public health services, and (7) benefits granted to war victims. We have discussed the nature and scope of social security systems, and we can now usefully attempt to discuss what social security statistics are, their sources and uses, and other related problems.

DEFINITION OF SOCIAL SECURITY STATISTICS

Social security statistics are defined as the statistics resulting from, and required for, the operation of a social security scheme. It is usual to collect the social security statistics[1] in addition to data on demographic, social and economic statistics, which are usually obtained from sources outside the scheme and are not usually considered as social security statistics.

In approaching the problems of compilation of social security statistics primarily from the point of the national system as a whole, we shall consider the sources of social security statistics, the types of schemes that should be regarded as part of a country's social security system, the elements of statistics in terms of which the operations of the various schemes should be measured, the uses of the social security statistics in the overall statistical system of the country, as well as in servicing the social security system itself, the limitations of the data, problems of compiling comprehensive national statistics and the question of international comparability.

SOURCES AND USEFULNESS OF SOCIAL SECURITY STATISTICS AND RECORDS

Sources

The sources of social security statistics depend in part on the nature of the organisation of the social security scheme, as well as the content of the scheme. The sources therefore vary according to the nature of the scheme.

(1) *Sources of Statistics Relating to Social Insurance*
The main sources of statistics on participants are the central registers maintained in connection with long-term insurance schemes and current records of the collection of contributions.

(2) *Sources of Statistics Relating to Public Service and Social Assistance*
No records or registers are usually kept; but in cases of care schemes, statistics of the medical care, both of services provided and of persons in receipt of benefits, can be obtained from periodical reports of doctors, hospitals and other medical institutions.

(3) *Sources of Social Security Records*
The records of social security systems constitute a valuable source of social and economic data, and these sources have the advantage that

they cover a wide range of human and economic endeavours and have a high degree of accuracy, due to constant checking and cross-checking and to the frequent revision of the basic data to cope with the administrative requirements of the social security system. While the record system varies considerably with the nature of the individual social-security system, certain broad features may be identified. In general, there are four basic elements in the records that relate to the insured population and the beneficiaries.

(a) *Register and individual records of insured persons*: This element deals with the name, sex, year of birth and/or year of entry into the scheme, marital status, registration area, and so on. This record provides a continuous history of the individual's insurable employment: that is, wages and/or payment of contributions.
(b) *Register of employers or establishments*: An essential element of social security records is the register of employers; this record may be used to identify separately the various establishments (public or limited company, partnership, location, industry, principal officers of the company, etc.) that may be operated by the employer. This register provides continuing data on the number of employees, the account of contributions paid, and fines or sanctions imposed for non-compliance with social security regulations.
(c) *Periodic employers' returns on contribution payments*: These consist of records of periodic returns of the employers on the payment of social security contributions by, or on behalf of, their employees. The report, which is usually of an aggregative character – number of employees, total contribution, and total or taxable payroll during the reporting period – also contains individual particulars of each insured employee: for example, wages, contributions, time worked.
(d) *Individual records of beneficiaries*: The records differ from country to country, and the system of record-keeping also differs considerably from one system to another.

There are a number of advantages and disadvantages that are related to these sources of records for purposes of statistical compilation and use. The individual records of insured persons constitute a valuable source of information for cross-sectional analysis according to various characteristics (e.g. age group, marital status) and for the compilation of work history of persons insured. A disadvantage is that they are not usually up to date, because the compilation takes time. The employer's register is generally more detailed and up to date. It provides a good frame for sampling employers and establishments and can be used to identify employers who are active and must send in periodic reports on employees. The periodic employer's report on contribution or payment is a key source of current statistical information. It lends itself to statistical compilation. The individual records of beneficiaries contribute to current statistics on unemployment, employment, injury, morbidity and health, pensioners and widows, and so on.

Coverage of Social Security Statistics

The social security statistics of a country normally cover the following types of schemes:

(1)schemes engaged in the following branches of social security: (a) medical care, (b) sickness benefit, (c) unemployment benefit, (d) old-age benefit, (e) employment injury benefit, (f) family benefit, (g) maternity benefit, (h) invalidity benefit, and (i) survivors' benefit; and
(2)schemes organised in accordance with the following principles: (a) social insurance (compulsory and voluntary), (b) public service, and (c) social assistance.

Statistical Information from Social Security Systems

Statistical information on the following items should be compiled periodically for each social security scheme:

(1)number of participants and beneficiaries;
(2)size of benefit (average and distribution); and
(3)statement of expenditure, income and assets of the institution.

Tabulation and Compilation of the Data

The feasibility of compiling social security statistics and providing a detailed breakdown of the data by various characteristics depends on the nature of the organisation of the basic records from which the statistical information is compiled. In general, however, it is useful to assign registration or identification numbers to each insured person and to each employer. This has the advantage of simplifying the statistical procedures and facilitates the compilation of data classified according to various characteristics.

USES OF SOCIAL SECURITY STATISTICS

Statistics concerning the operation of social security schemes serve as a basis for administrative decisions, for planning and for evaluating the role of social security in the social and economic organisation of the country. In general, adequate social-security statistics serve the following major objectives:

(1)to provide basic data for the administrative control of the social security scheme and for the appraisal of its operational efficiency;
(2)to provide a basis for the actuarial evaluation of the financial structure of the scheme, for the estimation of future financial requirements and for the making of short-term forecasts;
(3)to provide a means of appraising the social security programme as an instrument of social policy;
(4)to provide general information on social security for public use; and
(5)to supply data from administrative and accounts records to satisfy important needs in the fields of labour, economic, demographic and other statistics.

In addition to these uses, social security data can be put to use in the development of general purpose statistics in each country, but this will generally depend on the circumstances of each individual country, and account should be taken of the scope and character of the social security system and of the availability of statistics from alternative sources. The following uses, however, merit particular consideration:

(1)The data can be used as an indication of the number and characteristics of establishments;
(2)as a universe from which to draw samples;
(3)in the preparation of national accounts;
(4)as a source of information on employment and unemployment;
(5)as a source of information on morbidity;
(6)in the study of medical care directly administered under the social security scheme;
(7)in the compilation of statistics of industrial accidents and occupational diseases; and
(8)as a source of statistics on the labour force, wages, vital statistics and migration, subject to limitations of various kinds.

Limitations to the Wider Use of Social Security Data

The gap between the uses actually made of social security data and those which appear to be possible is due in particular to a number of limitations in the data. These are:

(1)inadequate recognition of the usefulness of the statistics;
(2)non-standard definitions and classifications;

(3)limitations due to coverage;
(4)time lag or delays in the publication of the statistics;
(5)the decentralisation of the records;
(6)changes in the scope of the data; and
(7)inaccessibility of records to other statistical agencies because of confidentiality regulations.

METHODS OF ANALYSIS OF SOCIAL SECURITY DATA

Statistics on the operation of social security systems are often more meaningful when correlated with other relevant social and economic data. Figures expressed in absolute terms may be of limited significance, but may provide the means for derivation of highly-significant relative measures or indices or indicators; for example, the average benefit for a given category of insured wage-earner expressed as a percentage of the average wage is such a relative measure. Relative measures not only are of value for purposes of evaluating the programme within a given country, but are generally much more meaningful for international comparisons than absolute figures, particularly monetary data: for example, social security expenditures expressed in national currency units. It is therefore useful, for the analysis of social security programmes, to develop relative measures, rates and indicators for the evaluation of the programme. The types of indicators and relative measures that can usefully be considered are now discussed:

(1)*Extent of social security coverage*: the number of persons insured or protected at a given date as a percentage of the corresponding total population, the number employed in a given industry, and so on.
(2)*Incidence and severity of contingency*:
(a) the number of new beneficiaries during a given period, usually a year, as a percentage of the average number of participants over the period;
(b) the average number of benefit days per case (e.g. per case terminated during a given period); and
(c) the number of benefit days during a given period divided by the average number of participants.
(3)*Old age indicator*: the number of old age pensioners as a percentage of the total population above the relevant minimum permissible age.
(4)*Level of contributions*: total amount of contributions as a percentage of the total wage bill for those covered.
(5)*Level of benefits*: the average benefit expressed as a percentage of

the average wage or income.

(6) *Economic indices of social security:*

(a) expenditure for social security as a percentage of national income or of consumption expenditure; and

(b) expenditure of public authorities for social security as a percentage of the total expenditure by public authorities.

(7) *Trends in real* per capita *expenditure for social security*: expenditure *per capita* throughout the years deflated by an appropriate price deflator.

CONCLUDING REMARKS

Social security statistics and the derived measures compiled at the national level provide the means for comparing the degree of social protection available in different countries. There are, however, a number of problems and limitations in all international comparisons of this nature. First, there are important differences in the national concepts and institutional arrangements for social security. No amount of statistical ingenuity can completely overcome such real differences. Second, the social security statistics of different countries are still subject to imperfections that impair the validity of the comparisons; for example, the concepts and definitions and methods underlying the various national statistics may fall far short of uniformity.

A common type of comparison of social security systems is one based on national figures expressed in monetary terms: for example, expenditure *per capita* expressed in national currency units. The conversion of national figures expressed in common monetary units (e.g. US dollars) introduces distortions. Subject to such important reservations, international comparisons based on social security statistics can yield useful results.

NOTE: CHAPTER 8

1 Social security statistics should be distinguished from social security records, which generate statistics or which may be used as a source of statistics for purposes other than the social security scheme itself.

REFERENCES AND FURTHER READING

Ghana (1969), *Statistical Handbook* (Accra: Central Bureau of Statistics).

Ghana (1969), *Directory of Industrial Establishments* (Accra: Central Bureau of Statistics).

Godfrey, V. N. (1974), 'A broader role for national provident funds: the Zambian experience', *International Labour Review*, pp. 137 – 52.
International Labour Organisation (ILO) (1961), 'Social security in Africa south of the Sahara', *International Labour Review*.
ILO (1964), *The Cost of Social Security, 1958 – 1960* (Geneva).
ILO (1969), *Social Security Planning and Implementation in Developing Countries* (Geneva).
ILO (undated), *Social Security Systems: A Report* (Geneva).
Mouton, P. (1975), *Social Security in Africa: Trends, Problems and Prospects* (Geneva: International Labour Organisation).
Ndive, M. O. (1976), 'Insurance statistics in Nigeria', PDS essay, Department of Economics, University of Ibadan, Ibadan.
Perrin, G. (1969), 'The future of social security', *International Social Security Review*, no. 1 (Geneva).
Tamburi, G. (1969), 'The ILO and social security: the challenge of technical co-operation', *International Social Security Review*, no. 4 (Geneva).
Vladimir Rys (1966), 'Comparative studies of social security: problems and perspectives', *Bulletin of International Social Security Association* (Geneva).
Vladimir Rys (1974), 'Problems of social security planning with particular reference to developing countries', Report No. ILO/TP/AFR/R17, International Labour Organisation, Geneva (mimeo.).

CHAPTER 9

Agricultural Statistics

Agriculture, apart from being an important mode of production, represents a life style in African countries. In African economies between 50 and 90 per cent of the population is engaged in agricultural activities (see, for example, Mouly and Costa, 1974); hence, the ability to support the population of African nations depends largely on their success in the agricultural sector. Agricultural statistics cover all activities of people engaged in the agricultural sector, and probably cover the largest area in African countries compared to any other sectoral statistics.

Agricultural statistics also play a special role in the assessment of the performance of African economies. The planning for full employment has to lean heavily on the agricultural sector statistics. The same is true of product-planning in these countries. Agricultural statistics also play a dominant role in estimating future production, income, consumption of immediate and final goods, and employment. Moreover, if agricultural production is studied in relation to final demand for goods, it serves as an indicator of inflationary trend in the economy.

Nor is this all. If we study the distribution of income in the agricultural sector in relation to that in the other sectors of the economy, we can get an idea of the increasing or decreasing disparities of income in this sector. When we analyse the data of prices of agricultural products in relation to the prices of the non-agricultural products, we may gain an insight into terms of trade between agriculture and the other sectors. Agricultural statistics are therefore important when they are studied independently and also when they are studied in relation to the other sectors.

This chapter describes the nature and scope of agricultural statistics, sources, limitations, and problems of interpretation and utilisation.

THE NATURE AND SCOPE OF AGRICULTURAL STATISTICS

The problem of collecting and estimating annual production of food crops and agricultural commodities is of great importance to all countries. It is more so with African countries, which are making strenuous efforts to feed their population and raise living standards. In addition to the collection of annual production figures for food crops, and so on, the following questions frequently arise in agricultural analysis: How many people work in agriculture? What is the number of their dependents? What is the relative importance of agriculture as a branch of industry from the employment point of view? What is the income per person in agriculture? What is the income per farm in different size classes of agricultural holdings, and how does agricultural work compare with the work in other branches of industry from the income point of view?

The information can be obtained in a variety of ways. The most important sources of such information in the past and maybe in the present as well, if the state of agricultural statistics in the world as a whole is taken into account, are population censuses and censuses of agriculture. The former provide data on agricultural populations that consist of (1) persons who are actually in agriculture, and (2) their dependents. The agricultural censuses offer data on farm populations: that is, all persons living on agricultural holdings. The 'agricultural population' as well as the farm population will show that part of the population responsible for agricultural production. It is important to note that if many from non-agricultural industries participate in the agricultural production through their secondary activities, or the agricultural population is active in non-agricultural industries, many analyses based on data about the agricultural population as identified in censuses of population may be grossly misleading.

SOURCES OF AGRICULTURAL STATISTICS

The principal sources of agricultural statistics in African countries are the marketing board statistics, rural household surveys, population censuses and the census of agricultural production. A detailed description of sources of agricultural statistics will be given for Nigeria. No attempt will be made to do the same for all African countries. Instead, sources for a few selected countries will be presented to give an idea of the variations and similarities between African countries in source material on agriculture.

In Nigeria, the statistics are compiled by the Federal Office of Statistics or Statistical Bureau, and the Ministry of Agriculture, and are published on a regular basis in the government publications. For instance, agricultural statistics can be obtained from the *Digest of Statistics* and *Economic Indicators*, published by the Federal Office of Statistics.

Sample surveys for estimating acreage planted, yield, population of domestic livestock, rural market prices and other items of information are carried out now in all the states of the Nigerian Federation by the Federal Office of Statistics on a regular annual basis. These surveys form the basis of the department's estimates of employment, production, consumption prices, and so on in the rural sector of the economy. This is in contrast to estimating production and other agricultural data through subjective appraisal by administrative and agricultural officers. The surveys cover the traditional sector of agriculture in Nigeria.

The scope of these surveys provides information on the following:

(1)area planted to crops and number of economic trees;
(2)average yields and production of crop items (cassava, cocoyam, etc.), which may remain in the ground for more than one crop year without being harvested;
(3)number of agricultural households and population of household members, showing age, sex and occupation;
(4)number of domestic livestock and poultry, showing breakdown by kind and sex;
(5)basic data on the farming pattern (use of fertilisers, agricultural implements used, transport facilities);
(6)crop calendar;
(7)rural market prices;
(8)prices received and paid by farmers; and
(9)household consumption (food and capital items).

The surveys are divided into three parts: livestock and poultry inquiry, farm surveys, and consumption surveys. The livestock and poultry inquiry is carried out three times in a survey year: in June, September and December. The inquiries in June obtain information on number, sex and kind of livestock kept by the households in the sample. The September and December inquiries collect additional information on livestock and poultry sold, born, dead, stolen or killed. This information is collected in order to estimate the reproductive and death rates of stocks and also to provide information on seasonal variations.

Farm surveys are designed to measure acreage planted to different crops and are weighed for recording purposes. The consumption inquiry takes place for a week at approximately monthly intervals over the survey year. It usually covers twenty households divided into two groups of ten households each, which are studied on alternate months and for a total of six weeks. Information is collected on consumption from own production, consumption from purchases (current) and other expenditure.

The sources that we have described may be termed 'traditional sources'. Apart from these sources, statisticians can use other documents such as records from official government publications, other governmental or non-governmental records of research institutes, individual studies, and so on to study the various aspects of agricultural development, or the component of agricultural growth, and to attempt to determine, on the basis of administrative or other documents, an estimate of the various characteristics of agriculture within the country.

We shall therefore attempt to provide a number of record sources for agriculture. These sources are indicative of what may be available, but they are by no means an exhaustive list of all the possible sources.

Statistical Yearbooks

The statistical yearbooks of individual states include topics on agricultural statistics. The same topics are covered by the *Digest of Statistics* published by the Federal Office of Statistics. This covers purchases of cocoa, palm oil, palm kernels, groundnuts, beniseed and seed cotton; and statistics on livestock (number of animals slaughtered on an annual basis as well as on a monthly basis for three current years prior to the date of publication of the *Digest*) are also provided.

Ministry of Agriculture and Natural Resources

(1) *Marketing Board Purchases*
The Produce Inspection Division usually has statistics for all crops sold to the marketing boards (groundnuts, cotton, cocoa, palm produce, soyabeans, sheanut) through the produce-buying agents. This information should be available for as far back as 1940, but in practice the information can only be provided from 1960 onwards for most states.

(2) *Crop and Weather Reports*
Each of the states is supposed to publish, on a regular basis, information on weather conditions and crop development and activities of the state Ministry of Agriculture and Natural Resources. Information given by division and the comparison of current prices with the previous year's prices. The state reports are drawn up from the reports of the divisions and therefore provide a summary of items covered in the division's report. This procedure may, in fact, suppress a lot of information. This publication is also classified as secret.

(3) *Livestock*
Quarterly reports of the Marketing Information Service of the Nigerian Livestock and Meat Authority from the states give figures on

cattle trade (local and imports figures), shipments to the south by different modes of transport, slaughter figures in the north, meat prices in the state capitals, and the prices of hides and skins.

Published Works on Agriculture

There are several articles and published works on agriculture in Nigeria. One of the most important publications is *Agricultural Development in Nigeria, 1973 – 1985*, published by the Federal Ministry of Agriculture and Natural Resources. It gives fairly comprehensive statistics on Nigerian agriculture over the years in the form of time series data, which may be useful for further statistical analysis.

Writing on the gaps in the existing sources of statistics on agriculture in Nigeria, Enabor (1976) has commented:

> There are no published data on production and consumption of wood and wood based products . . . The major source of primary data on wood products is FAO's Country Tables of Production, Trade and Consumption of Forest Products: Africa, 1960 – 1973, published in 1975. FAO's data are compiled from information supplied by individual reporting countries. The organisation, however, usually edits the country's data and sometimes inserts its own estimates for the year(s) that a particular country has failed to supply required information. There is, therefore, some broad correspondence between FAO's and individual countries' estimates of production, trade and consumption of wood products. Nevertheless, the published statistics of wood products in Nigeria are entirely inadequate and unsatisfactory. The government should endeavour to collect and collate comprehensive statistics on national production, trade and consumption of all forest products and ensure their timely publication.

The review of sources of agricultural statistics has dealt with Nigeria. A similar review for a few other African countries in in order. In Kenya, the main source of agricultural data is the *Statistical Abstract*, published by the Kenya Ministry of Economic Planning and Development. Agricultural statistics can also be obtained from the *Statistical Digest* and *Economic Survey*. In Tanzania, the sources of agricultural statistics are the *Economic Survey*, published annually, and the *Budget Survey*. In the case of Uganda, the *Statistical Abstract*, published by the Ministry of Planning, the annual reports of marketing boards and the *Quarterly Digest of Statistics* are the principal sources of agricultural statistics.

In Mauritius, the chief sources of data include the *Bi-Annual Digest of Statistics*, published by the Central Statistical Office, the *Annual Report* of the Department of Agriculture and the *President's Report*, published annually by the Mauritius Chamber of Agriculture. For Malawi, the basic source is the *Economic Report*, published by the Office of the President and Farmers' Marketing Board. The main sources of the Zambian agricultural statistics include the *Monthly Digest of Statistics*, published by the Central Statistical Office, and the *Economic Report*, published annually by the Ministry of Development and Finance. Also relevant are the publications of the National Agricultural Marketing Board, the Cold Storage Board and the Dairy Produce Board – the three principal agricultural marketing boards in Zambia.

The above analysis of the sources of statistics clearly shows that the agricultural data are collected by more than one agency in a country, and that in a number of countries data on agriculture are published on a quarterly and monthly basis, while in other countries such data are available only annually.

METHODS OF ANALYSIS

As with the sources, the computation of agricultural statistics varies from one African country to another. In Algeria, Chad, Madagascar, Malawi, Morocco, Nigeria and Upper Volta, estimates of production for the various kinds of cereals are made by multiplying the estimated average yield per unit of area by the corresponding area under each kind of cereal.

In Ethiopia, direct information is usually used for making estimates of the total production of cereals. This direct information on production is obtained through crop-reporters. However, no single method is adequate for estimating agricultural output in African countries. Output of agriculture has to be estimated directly in some cases and indirectly in others. For instance, in Nigeria, agricultural crops are divided into three categories, and different methods and sources are used to estimate the output of each item.

The first category consists of the crops grown mainly for export. These crops are cocoa, palm kernel, cotton and rubber. For cocoa, the marketing board purchases are used to estimate the output; for palm kernel, a combination of information of purchases of palm kernel by the marketing boards and the inputs of fruits in the milling of oil supplied by the industrial survey has been used. Producers' consumption of palm kernel has been neglected. In the case of cotton, sample survey figures have been used. The information on the production of rubber, which is partly processed in the country and partly exported, has been drawn from the industrial survey results (rubber input of the processing establishments) and export statistics. The second category consists of the crops grown mainly for domestic use. This is further divided into farm crops, beans and plantain. The sample survey results are taken with regard to the area and the yield. In the case of the beans and plantain, export data supplemented with estimates of domestic consumption of these items are used. Estimates of output of the third category (other crops) are based on the exports only.

Similarly, in all other countries of Africa, household sample surveys[1] and the marketing boards statistics play a crucial role in supplying information on the total agricultural output.

CONCEPTUAL AND MEASUREMENT PROBLEMS

So far we have been concentrating on the sources and methods of compiling statistics in agriculture. But we have not tackled the various problems encountered when contemplating conducting an agricultural survey. These problems are of a conceptual, definitional and practical nature. Definitional problems relate to the concepts employed in collecting the data. In 1969 a report of the World Census of Agriculture taken in 1960 was published by the Food and Agricultural Organisation (FAO), and this report pinpointed a number of these problems.

Conceptual Problems

First, in dealing with the conceptual problems, substantial adjustments had to be made to adapt the programme for the 1960 World Census of Agriculture to the special pattern of agriculture and social conditions prevalent in Africa. The main adjustments are due to a number of factors.

First, under traditional African conditions, it is not always clear what is to be considered the main occupation of the holder, as persons can be occupied in different occupations at different times of the year. Egero (1973), describing this problem, has written on Tanzania:

> Household data on agricultural activities proved quite difficult to interpret, partly due to the uncertainties as to how polygamous households have been treated, but mainly related to problems in the translation. Firstly, the distinction between what was called 'main' and 'secondary' agricultural produce was never really clear; secondly, these headings were both translated into plural form in Swahili with the result that rarely as few as two crops appeared on

the questionnaire. Arbitrarily, the two first recorded crops were chosen for coding in the processing of the data.

On the difficulties of measuring the agricultural population,[2] Koley (1973) has remarked:

> It is, however, to be mentioned here, that the data of economically active women in the country are extremely unreliable due to the confusion arising from the definition of economic activity and hence uniformity in definition could not be maintained throughout the country. The number of employees (for both males and females) may be regarded as reliable, but a lot of confusion arose in classifying people, particularly women, into own-account or family workers. This was due to the fact that in quite a few places a man can have more than one wife who independently cultivate separate shambas though the shamba may customarily belong to the head of household. Hence, it is not surprising that almost the entire economically active female population barring a few in urban areas are engaged in agriculture either as own-account or as family workers. Among economically active males also, agriculture is by far the most predominant occupation, keeping 85% of the population engaged. As a contrast, farm employment is dominated by males; among employers 77% are male and among employees 95%.

To solve such problems, the African programme provided alternative items, namely: (1) agricultural occupation, and (2) agricultural and other occupation.

Second, in certain countries of the region, farmers frequently work away from their holdings during a large part of the year: for example, on plantations, in mines or factories. Their own smallholdings are then operated mostly by other members of the family. In order to compare the output of such holdings with those under the continuous supervision of the holder, the regional programme suggested that the information should be classified as follows: (1) holdings under the continuous supervision of the holder, and (2) holdings where the holder worked away from the holding during a continuous part of the year.

Third, under the legal status of the holder, the African programme provided for recording the cases where the holding is operated by a joint and extended family, as distinct from those operated by an individual holder. Furthermore, in order to assess the progress of the work undertaken in agriculture, the regional programme required separate information on (1) holdings operated under a definite plan, and (2) holdings operated under the traditional system.

Fourth, communal grazing land is of particular importance in Africa. A large proportion of agricultural production is derived from livestock grazed on communal land, while production obtained from arable land is of limited importance. The picture on agriculture would be incomplete without due regard to communal grazing land in censuses. The African programme thus added two questions: (1) 'Does the holding consist of communal grazing land?' (2) 'If not, has it access to communal grazing land?' In this way the relationship between individual holding and communal land could be clearly seen. The programme further suggested that countries concerned might also collect additional information relating to the number of livestock and the number of months grazed on communal land. Area used under mortgage was also listed in the African programme, as in the Asian adaptation.

Fifth, in the case where permanent and temporary crops are grown simultaneously in the same field, the allocation of the land into the broad classes of land utilisation is extremely difficult under prevailing African conditions. Each field recorded only once could be allocated to the class of crops considered most important. An excellent example on the difficulties of estimating crop acreage comes from Tanzania, as Koley (1973) has observed:

> It may be stressed here that yield figures are in most cases subjective estimates without specifying yield for mixed croppings and hence are subject to considerable margins of error. Hence, statistics derived from these acreage figures should be treated with caution . . . A little adjustment has been necessary for the derivation of coffee acreage and production, more than a fifth of which is grown on estates. Two important crops, namely cassava and bananas, have been left out due to difficulties in obtaining reliable yield figures. It is worth noting that cash crops like cotton, coffee, and cashewnuts have, on an average, less acreage per household than food crops, like maize, sorghum, etc. This could be due to the reasons that whereas the cash crops are normally grown pure, food crops are mostly grown in mixtures.

The compromise adopted in the African programme was to subdivide land under temporary crops (as well as land under permanent crops) into (1) land exclusively under temporary/permanent crops, and (2) land mainly under temporary/permanent crops.

Sixth, it was realised that the African programme asked for irrigated and non-irrigated area under arable land – land under permanent crops and for cultivated meadows and pastures. In addition to the total area under production, the African programme proposed that, in the case of mixed and associated crops, the following areas should be reported under individual crops: (1) area of the crop grown as a single crop, and (2) area of the crop grown as an associated crop. This breakdown was suggested in view of the economic importance attached to the extremely complicated question of mixed and associated crops in this region. In addition, a new table was suggested in which countries were asked to list the main crop mixtures and associations, giving the area and the number of holdings reported.

The world programme proposed that successive surveys should collect information on employment in agriculture. Since under African conditions this was not feasible, the regional programme suggested that, in addition to the number of persons employed during the census week, the maximum number of persons employed during the census year should be investigated. Because of the obvious duplications involved, it is of interest to understand the labour problems in homogenous areas where the peak period for the different holders falls at the same time. Countries that found it difficult to investigate employment in agriculture could still obtain valuable information from the holder as to whether he had used paid labour during the preceding agricultural year.

The regional programme also suggested the investigation of the problem of migrant labour in countries where it is important.

The major change with regard to the farm population was the adoption of the concept of household to African conditions. The housing unit in Africa was found in many cases to be composed of a group of huts, sometimes referred to as a compound, where in a polygamous society the husband lives in one hut and the other huts are occupied by his wives and children. These people do not live under the same roof, nor do they always take their meals together, as required by the definition of a household in the world programme. The criteria also do not apply in cases where the fields away from the village are operated by the members of the household, living in small camps near the field for the crop seasons, or even for the whole year in the case of permanent crops. In these circumstances, the criterion that the members of the household should share the community life was considered to be an essential element on which the definition of household could be based. Accordingly, the household for the purposes of enumerating the farm population in Africa was defined as the aggregate of persons generally bound by the ties of kinship, who normally reside together, not necessarily under the same roof.

In certain African countries, cultivation for crops is made possible only after the river water has flooded the area and then receded. In order to obtain information on the extent of this practice, the African programme included the question: 'Of the land actually irrigated, how much was flooded by rivers?'

Practical Problems

The second difficulty relates to practical problems. Apart from not being able to apply the known concepts to African countries, there are practical problems faced by these countries while attempting to collect data by surveys.

Lack of trained personnel is a major problem.

Another problem is the lack of communication. Distances between villages are long, and the villages are sparsely distributed. Road communication is extremely poor. Accordingly, to reach the respondent presents a serious problem. An added difficulty is that, even if enumerators can contact respondents, they may not be able to get the required information because of illiteracy, superstition and unwillingness on the part of the respondents to supply any information.

Data-processing of the results is also usually very slow.

On the general reliability of the data from such agricultural surveys, Koley (1973) has observed that:

> But the major limitations in ascertaining the accuracy of the data lie not in sampling but in reliability of the source of information, i.e. the respondents themselves as well as the enumerators sometimes. The definitions such as 'economic status', 'main and secondary agricultural produce' were difficult for the census experts to convey properly to the enumerators, not to mention the difficulties in extracting the exact information from the normally reticent respondents. Hence, the statistics presented in this chapter should be handled with great caution. An attempt has been made to compare some of the findings of the census against similar information available from other sources such as the Ministry of Agriculture. However, this sort of comparison has its obvious limitations due to the very nature of the data.

All these conceptual and measurement problems in agricultural statistics have to be borne in mind when using the data. Together with these limitations, we have to face additional problems when we want to use published statistics in government publications or want to use different sources to build our own series of output.

OTHER DIFFICULTIES IN COLLECTING AGRICULTURAL STATISTICS

There are also a number of other difficulties connected with the collection of agricultural statistics in Africa. These can be classified into four categories: (1) problems of confidentiality, (2) problems of coverage, (3) problems of timeliness, and (4) problems of comparability.

Problems of Confidentiality

The returns are used for official purposes; and if an outsider, as is the case of researchers, wants to see the individual returns in order to remove inconsistencies from his results, or doubts the accuracy of the data, he is promptly told: 'Sorry, this is confidential.' The confidentiality of the individual returns is provided by law, and a statement to this effect is made on the questionnaire itself. However, these legal provisions are applied somewhat indiscriminately in many African countries. Almost all data are labelled confidential, even data on crops and weather.

Problems of Coverage

Even the limited data that can be gleaned from published statistics or records are extremely uneven in coverage and quality. The data on agricultural production are generally under-reported for want of coverage. The magnitude of this under-reporting is even hard to guess. The problem is aggravated when one wants to use the data compiled by different agencies. A researcher is often confronted with the problem of synchronising different sources of data on the same item produced by different statistical agencies. These differences in data stem, apart from other things, from differences in coverage.

Problems of Timeliness

Most of the statistics from these records are either reported late or not reported at all. Obviously, the former situation is preferable to the latter, although none is desirable. An argument often advanced in favour of the late publication is that it improves the quality of statistics. But the timeliness of statistical information can be increased only at the expense of accuracy, while improvements in quality require more time and increase the cost of the information. Unfortunately, in Africa, the late publication of statistics is coupled with unreliability also. It cannot therefore be overstressed that, in the event of late publication, steps should be taken to ensure that the original data do not get lost before the analysis and publication. Furthermore, the slack operation of the system of data collection leads to non-collation or defective collation of information, which may be irretrievable when the faults are detected.

Problems of Comparability

Comparability of the agricultural statistics in Africa, both over time (temporal comparisons) and over space (spatial comparisons), is limited by the use of non-comparable definitions, difference in coverage and the timing of the data collection. Comparability is further reduced by the details of tabulations. The remedy for this is that procedures for the recording of information and tabulation should be uniformly applied for all states. This can be achieved through co-ordination of the activities of the statistical staff through seminars and workshops.

CONCLUDING REMARKS

Running back over the deficiencies and gaps that we have mentioned, it is possible to break down the problem into specific areas. First, the nature of agricultural statistics is such that they have a lower forecasting ability than statistics from other sectors of the economy, because of the great influence of the weather on production; this means that they lose a lot of their usefulness by late publication. Of course, complaints about delays in publishing statistics are an occupational hazard of the trade.

Second, the dispersed nature of observation units also imposes additional problems; the extra difficulties in data collection are compounded because there are numerous and varied transactions to be recorded.

Third, the statistics suffer from gross under-recording of small farms' output and the output that is consumed by the producers (farmers) themselves, and the proportion of it that is bartered. The significance of these items can hardly be overemphasised in the semi-monetised economics, and particularly in the rural sectors of African countries.

Fourth, concepts and definitions devised for administrative schemes may not suit statistical needs, and the information may not cover the whole of the relevant population. Many of the records may also be out of date, particularly where individuals have only infrequent and irregular contact with the administrative systems.

Finally, data on agriculture (with the exception of many subsistence activities) are already more plentiful than those on many other economic activities. In our view, they therefore do not require as high a priority as those concerning some other spheres, such as small-scale urban activities. It may, however, be useful if the agricultural sample surveys produce more data on incomes, expenditure and other household characteristics and sample a wider range of rural situations.

Changes of method in small farm surveys, so that they gave additional data on labour inputs by sex and age in relation to household structure, would also be desirable. Resources permitting, in the long run it would also be useful to aim at obtaining information on the time that women spend on the preparation of food and other non-agricultural tasks.

NOTES: CHAPTER 9

1 For further discussion of surveys conducted in African countries, see, for example, Economic Commission for Africa (1970) pp. 95 – 111.

2 The agricultural population is almost entirely rural, although there are traces of agricultural activity in urban areas also.

REFERENCES AND FURTHER READING

Abdulahi, H. (1972), *Population and Livestock Census of the Governorate General of Arussi*, Staff Report No. 1 (Addis Ababa: Central Statistical Office).

Abraham, W. I. (1972), *Ethiopia's Estimates of Cereal Production: How Good Are They?*, Staff Report No. 3 (Addis Ababa: Central Statistical Office).

Blanchard, D. (1971), *Some Problems of Increasing Output with Peasant Production Units* (Legon: Department of Economics, University of Ghana).

Bondestam, L. (1973), *Some Notes on African Statistics: Collection, Reliability and Interpretation*, Research Report No. 18 (Uppsala: Scandinavian Institute of African Studies).

Boserup, E. (1965), *The Conditions of Agricultural Growth: The Economy of Agrarian Change under Population Pressure* (Chicago, Ill.: Aldine Publishing).

Brown, T. M. (1972), 'Macroeconomic data of Ghana', *Economic Bulletin of Ghana*, vols 1 and 2.

Buxton, H. J. (1973), 'The cash income of cocoa farmers in Ghana', paper presented to the Cocoa Economics Research Conference, University of Ghana, Legon.

Due, J. M. (1969), 'Agricultural development in the Ivory Coast and Ghana', *Journal of Modern African Studies*, no. 4.

Dutta-Roy, D. K. (1969), *The Eastern Region Household Budget Survey*, (Legon: Institute of Statistical, Social and Economic Research, University of Ghana).

Economic Commission for Africa (1970), *Survey of Economic Conditions in Africa*, Vol. 2, (Addis Ababa).

Egero, B. (1973), 'Non-sampling errors', in Egero and Henin (1973).

Egero, B. and Henin, R. A. (eds) (1973), *The Population of Tanzania: An Analysis of the 1967 Population Census*, Census Vol. 6 (Dar es Salaam: Bureau of Statistics).

Eichberger (1968), 'Food production and utilization in Ethiopia in 1958', Addis Ababa (mimeo.).

Enabor, E. E. (1976), 'World consumption requirements in Nigeria: reassessment', *Nigerian Journal of Economic and Social Studies*, vol. 18, no. 1, pp. 121 – 46.

Ewusi, K. (1971), *The Distribution of Monetary Incomes in Ghana* (Legon: Institute of Statistical, Social and Economic Research, University of Ghana).

Food and Agriculture Organisation (FAO) (1969), *Report on the 1960 World Census of Agriculture*, Vol. 2 (Rome).

FAO (1973), *Forest Products Statistics Review, 1961 – 72* (Rome).

FAO (1975), *Country Tables of Production, Trade and Consumption of Forest Products: Africa, 1960 – 73* (Rome).

Hart, K. (1973), 'Informal income opportunities and urban employment in Ghana', *Journal of Modern African Studies*, no. 1.

Holmberg, J. (1972), 'The reliability of estimates from surveys of Ethiopian peasant farmers', Central Statistical Office, Addis Ababa (draft).

Jabati, S. A. (1967), 'Agricultural development in Sierra Leone', *Sierra Leone Trade Journal*, 7, pp. 7 – 17.

Janelid, I. (1971), 'The socio-economic role of farm women in agricultural and national development in West Africa', paper prepared for FAO conference on Agricultural Education at the Intermediate Level, Zaria.

Kenya (annual), *Statistical Abstract* (Nairobi: Economics and Statistics Division, Ministry of Economic Planning and Development).

King, R. P. (1977), *An Analysis of Rural Consumption Patterns in Sierra Leone and their Employment and Growth Effects*, African Rural Employment/Economy Working Paper No. 21 (East Lansing, Mich.: Department of Agricultural Economics, Michigan State University).

Knight, J. B. (1972), 'Mechanism of income distribution in Ghana' (draft, mimeo.).

Knight, J. B. (1972) Rural-Urban Income, Comparisons and Migration in Ghana, *Bulletin of the Oxford Institute of Economics and Statistics*, No. 2.

Koley, C. (1973), 'Agricultural data', in Egero and Henin (1973).

Lewis, R. (1953), 'Agriculture and planning in Sierra Leone', *Corona*, 5, pp. 209 – 12.

Makennen, A. (1971), *Cotton and its Significance in the Ethiopian Economy* (Addis Ababa: Department of Economics, Haille Selassie University).

Mauritius (bi-annual), *Bi-annual Digest of Statistics* (Rose Hill: Central Statistical Office).

May-Parker, I. I. (1973), 'Agriculture and food production', paper presented at the conference of the Planned Parenthood Association of Sierra Leone, Freetown.

Mensah, J. H. (1971), 'Incomes distribution in Ghana', *Business Weekly* (Accra, May).
Mouly, J. and Costa, E. (1974), *Employment Policies in the Developing Countries* (London: Allen & Unwin).
Nigeria (1972), *Rural Economic Survey of Nigeria* (Lagos: Federal Office of Statistics).
Nigeria (annual), *Digest of Statistics* (Lagos: Federal Office of Statistics).
Nigeria (undated) *Agricultural Development in Nigeria, 1973 – 1985* (Lagos: Federal Ministry of Agriculture and Natural Resources).
Njoku, A. O. (1971), 'Labour utilization in traditional agriculture: the case of Sierra Leone farms', unpublished Ph.D. thesis, University of Illinois, Urbana.
Ojo, M. O. (1977), 'Food supply in Nigeria, 1960 – 1975', *Economic and Financial Review*, 15 (Lagos: Central Bank of Nigeria).
Ord, H. W. *et al.* (1964), 'Projected level of demand, supply and imports of agricultural products in 1965, 1970 and 1975', report prepared for US Department of Agriculture, Edinburgh.
Reusse, E. (1968), 'Ghana's food industries, 1968', Food and Agriculture Organisation, Rome (mimeo.).
Rourke, B. E. (1973), *The Value of Agricultural Production by Region in Ghana, 1970 – 1: Some Provisional Estimates* (Legon: Institute of Statistical, Social and Economic Research, University of Ghana).
Rourke, B. E. and Obeng, F. P. (1973), 'Seasonality in the employment of casual agricultural labour in Ghana', *Economic Bulletin of Ghana*, no. 3.
Rourke, B. E. and Sakyi-Gyinae, J. K. (1972), 'Agricultural and urban wage rates in Ghana', *Economic Bulletin of Ghana*, no. 1.
Sierra Leone (1961), 'Development plan of the agricultural services of Sierra Leone', Department of Agriculture, Freetown (mimeo.).
Sierra Leone (1967), *Agricultural Statistics Survey of Sierra Leone, 1965 – 1966* (Freetown: Central Statistics Office).
Snyder, D. W. (1971), 'An econometric analysis of household consumption and savings in Sierra Leone', unpublished Ph.D. dissertation, Pennsylvania State University, USA.
Spencer, D. S. C. and Byerlee, D. (1977), *Small Farmers in West Africa: A Descriptive Analysis of Employment, Incomes and Productivity in Sierra Leone*, African Rural Economy Working Paper No. 19 (East Lansing, Mich.: Department of Agricultural Economics, Michigan State University).
Stoces, F. (1966), 'Agricultural production in Ghana', *Economic Bulletin of Ghana*, no. 3.
Uganda (annual), *Digest of Statistics* (Entebbe: Statistics Division, Ministry of Planning).
Uganda (annual), *Statistical Abstract* (Entebbe: Statistics Division, Ministry of Planning).
Zambia (monthly), *Monthly Digest of Statistics* (Lusaka: Central Statistical Office).

Statistical digests and statistical abstracts of African countries (see Chapter 1).

MAJOR SOURCES OF AGRICULTURAL STATISTICS

ALGERIA

Revue statistique agricole, nos 8 and 9 (1969).

BOTSWANA

Department of Agriculture, *Agricultural Survey* (1967 – 8, 1968 – 9; annual).

Division of Planning and Statistics, *Agricultural Survey* (1969 – 70, 1970 – 1).

Central Statistics Office, *Freehold Farm Survey* (1970 – 1; annual).

Meat Commission, *Report and Accounts* (1967, 1968, 1969; annual).

BURUNDI

Ministère de l'Agriculture et de l'Elevage *Burundi agricole* (1967, 1968, 1969, 1970, 1971).

CAMEROUN

Development Corporation, *Annual Report* (1967, 1968, 1969).

Direction de l'Agriculture, *Statistiques agricoles Est Cameroun* (Yaounde, 1967 – 8).

CHAD

Ministère de l'Agriculture et de la Production Animale, *Rapport annuel* (1967).

Ministère de l'Elevage, *Rapport annuel* (1967, 1968, 1969).

ETHIOPIA

Department of Land Tenure, Ministry of Land Reform and Administration, *Annual Report* (1967 – 8, 1968 – 9).

National Coffee Board, *Coffee Statistics Handbook* (1970 – 1; annual).

National Coffee Board, *Ethiopia Coffee Review* (1972, 1973).

Institute of Agricultural Research, *Report for the Period 1966 – 8, 1968 – 9* (annual).

Livestock and Meat Board, News Digest (1968; bi-monthly).

Statistical Bulletin, no. 3: *Major Crops Seasons in 13 Provinces* (Addis Ababa, July 1970).

GHANA

Cocoa Marketing Board, *A Quarterly Magazine* (1967, 1968, 1969, 1970, 1971).

Fisheries Department, *Fish Quarterly Bulletin* (1968, 1969).

IVORY COAST

Chambre d'Agriculture de la Côte d'Ivoire, *Bulletin mensuel* (1970, 1971).

KENYA

Cotton Lint and Seed, *Annual Report and Accounts* (1968, 1969, 1970).

Kenya Tea Development, *Annual Report and Accounts* (1967 – 8, 1968 – 9).

Ministry of Agriculture, *Quarterly Economic Review of Agriculture* (1969, 1970, 1971, 1972).

Dairy Board, *Annual Report* (1966 – 7, 1968 – 9).

Land and Agriculture Bank of Kenya, *Annual Report* (1969).

Kenya Sisal Board, *Bulletin* (1970, 1971; quarterly).

Economics and Statistics Division, Ministry of Economic Planning and Development:
Economic Survey (annual).
Statistical Digest.
Farm Economic Survey Report, no.28 (1972), no. 27 (1964 – 6 to 1967 – 8).

Fishery Division, *Report on Kenya Fisheries* (1967 – 8; annual).

Kenya Wheat Board, *Report* (1967 – 8, 1968 – 9, 1969 – 70, 1970 – 1).

LESOTHO

Ministry of Agriculture, Co-operatives and Marketing, *Annual Report* (1969, 1970).

LIBERIA

Department of Agriculture, *Annual Report* (1969 – 70).

MADAGASCAR

Ministère de l'Agriculture, de l'Expansion Rurale et du Ravitaillement, *Rapport annuel* (1968 – 9).

MALAWI

National Statistical Office, *Compendium of Agricultural Statistics* (1971) (annual).

Department of Forestry and Game, Forest Research Department 6, *Report*.

Office of the President and Farmers' Marketing Board, *Economic Report* (annual).

MALI

Service de la Statistique Generale, *Rapport definitif de l'enquête agricole* (1968 – 9).

MAURITIUS

Forest Department, *Annual Report* (1967).

Forestry Service, *Annual Report* (1968, 1969, 1970).

Central Statistical Office, *Bi-Annual Digest of Statistics*.

Department of Agriculture, *Annual Report*.

Chamber of Agriculture, *President's Report* (annual).

NIGERIA

Federal Office of Statistics, *Economic Indicators* (monthly).

Federal Office of Statistics, *Digest of Agricultural Statistics: Western Nigeria* (1970; annual).

Cocoa Research Institute, *Annual Report* (1967 – 8, 1968 – 9, 1969 – 70, 1970 – 1).

Federal Department of Fisheries (Lake Chad Research Station), *Annual Report* (1966 – 7, 1969).

Fisheries Service, *Annual Report* (1967, 1968, 1969).

Department of Forest Research, *Programme of Work* (1970 – 1, 1971 – 2).

Agricultural Credit Corporation, *Western Annual Report* (Ibadan, 1966 – 7, 1967 – 8, 1968 – 9).

Forestry Department, *Western Annual Report on the Forest Administration* (1966 – 7, 1967 – 8).

SIERRA LEONE

Survey and Land Department, *Report* (1966 – 7, 1967 – 8, 1968 – 9, 1969 – 70; annual).

SWAZILAND

Sample Census of Agriculture, pt 1, (March and June 1971).

Census of Individual Tenure Farms, 1970 – 1 (June 1972).

Report on Commercial Timber Plantation and Wood Products Statistics (1971, 1972).

Ministry of Agriculture, *Annual Report* (1967, 1968, 1969, 1970, 1971).

TANZANIA

Fisheries, *Annual Report* (1967, 1968).

Forest Division, *Annual Report* (1967, 1968).

Central Statistical Division, *Economic Survey* (annual).

Central Statistical Division, *Budget Survey*.

TUNISIA

Secrétariat d'Etat au Plan, *Enquête agricole annuelle*:
Bizerte (March – April 1969), Gabes (April 1969), Jendouba (April 1969), Nabeul (March 1969), Tunis (April 1969), Beza (March – April 1969).

Secrétariat d'Etat au Plan, *Statistiques agricoles* (1969; monthly).

Institut de Statistiques, *Statistiques agricoles trimestrielles* (1969).

UGANDA

Ministry of Agriculture, *Agricultural Newsletter* (1967, 1968, 1969, 1970; bi-monthly).

Department of Agriculture, *Report on Annual Agriculture Statistics* (1967 – 8).

Statistics Division, Ministry of Planning and Economic Development (Entebbe), *Quarterly Digest of Statistics*.

ZAIRE

Ministère de l'Agriculture, *Revue d'information agricole* (1967, 1968, 1969, 1970; bi-monthly).

ZAMBIA

Department of Wildlife, Fisheries and National Ports, *Annual Report* (1969, 1970).

Forest Department, *Annual Report* (1967, 1968, 1969, 1970).

Dairy Product Board, *Annual Report and Accounts* (1967 – 8, 1968 – 9).

Ministry of Development and Finance, *Annual Report*.

National Agricultural Marketing Board, various publications.

Cold Storage Board, various publications.

Dairy Produce Board, various publications.

CHAPTER 10

Financial Statistics

The last twenty or so years have witnessed a steady improvement in the statistical coverage, collection, analysis, publication and uses of financial and monetary statistics in Africa. There has also been a comparable increase in the range of statistics available, particularly on financial assets and liabilities. These statistics have been brought together and published with considerable regard to consistency and comprehensiveness.

These developments incidentally coincided with the establishment of central, commercial and development banks by many African countries at the dawn of independence. The monetary authorities and international organisations such as the World Bank (or International Bank for Reconstruction and Development, IBRD) and the International Monetary Fund (IMF) have also played a considerable part in these developments. Nevertheless, users of financial and monetary statistics will find that many gaps and inadequacies still exist in the financial and monetary statistics collected and published in Africa. This chapter will describe the nature of financial statistics, sources, general arrangements for collecting such statistics, their uses and the limitations of the statistics.

NATURE AND SCOPE OF FINANCIAL STATISTICS

We begin our discussion by explaining the nature of financial statistics. Let us start with the production process at a particular point of time with a given amount of factors of production. These factors of production jointly produce certain levels of goods and services. If these factors of productions are paid in terms of the goods that they themselves produce, or in terms of other goods that they require for consumption, the payment will be called non-monetary. For instance, if a farmer employs some labour on his farm and as a reward pays in terms of rice, no use of money has been made in the transaction, and hence it will be called a non-monetary transaction.

Non-monetary transactions find their origin in barter also. In the initial agrarian societies, the daily needs were met by exchanging commodities for commodities. If a farmer requires cloth, he can trade his produce with that of the weaver. There are obvious difficulties in the barter system of finding a customer so as to coincide the needs and also of determining the terms of trade.

These difficulties led to the introduction of money as a medium of exchange. Money covers most of the transactions in present-day developing countries. Money includes paper currencies and deposits with the banking system. In the present day economies, the functions of money are extended from being only the medium of exchange to serve as a store of values, a measure of values and many other functions. When we buy things from a market, we pay in terms of money, and money is used in most of the transactions.

The existence of money has led to the development of the banking system. These banks are called financial intermediaries, because they issue loans against deposits made by other persons. If a man requires some money to carry out his investment plan, instead of borrowing from other persons, he can directly borrow from the banks where these persons have deposited their money. Thus, banks simply transfer the savings of one person or a group of persons to the other person or a group of persons. There are different types of banks: commercial banks, central banks, mortgage banks, savings banks, co-operative banks, investment trusts, and so on. All these banks are called financial intermediaries. The central bank of a country regulates the operations of other banks by adopting several measures. The transactions that involve the use of money and credit are termed monetary transactions.

These monetary transactions can be subdivided into the financial monetary transactions and the non-financial monetary transactions. The financial monetary transactions relate to paper claims issued by the financial intermediaries and the producing units in the form of bills of exchange, promissory notes, shares and debentures, and so on. The owner of these claims commands a claim on the real resources on the specified date of maturity and for the specified amount. The financial transactions relate to the borrowing and debt repayments and accumulation and reduction in financial assets.

The non-financial transactions have their origin in the productive process. When a factor of production is paid a reward in terms of money for its contribution to production in the form of wages, profits, rent or interest, it is termed as non-financial. Further, out of the income, taxes are paid, and gifts and donations are made. These transactions are also called non-financial transactions, as they do not relate to the incurring of liabilities but add to the financial wealth. Thus, the transactions relating to the earning of income and unilateral transfers are called non-financial transactions. Financial statistics cover all monetary transactions, whether they are financial or non-financial, and ignore the non-monetary ones.

Basically, the monetary transactions can be grouped as in Table 10.1

Table 10.1 *Monetary transactions*

Non-financial transactions	*Financial transactions*
Payroll (wages and salaries)	Currency and deposits
Interest	Bank loans and mortgages
Rents and royalties	Corporate securities
Dividends	Provident funds
Taxes and refunds	Life insurance premiums and claims and the like
Gifts and donations and the like	

The financial statistics cover the financial transactions. But since all monetary transactions are carried out by all sectors of the economy through financial intermediaries, the financial statistics records cover all the sources and uses of funds (finance). The sources of funds relate to the decrease in assets and/or increase in financial liabilities, while uses of funds relate to the increase in assets and/or decrease in liabilities. The financial institutions, to which the financial statistics relate, are: monetary authority (central bank), commercial banks, clearing houses, acceptance houses, money and capital markets.

SOURCES OF FINANCIAL STATISTICS

Financial statistics are collected and published both at the national level and at the international level. Each country's central bank has been assigned the responsibility of collecting data from all the financial institutions in the country and publishing these data. The financial institutions in their turn also bring out publications showing their statement of accounts and the participation in economic activities. At the international level, the IMF publishes, on a monthly basis, *International Financial Statistics*. Similarly, the agencies of the World Bank publish the statement of the loans granted to various developing countries for carrying out their development programmes.

National Sources

First, we deal with the national sources of financial statistics. Starting with Nigeria, the financial statistics are published in the four publications of the Central Bank of Nigeria. These are: *Annual Report and the Statement of Accounts*, *Economic and Financial*

Review, *Monthly Report* and *Development in the Nigerian Economy in the First Half of the Year.*

The *Annual Report and the Statement of Accounts*, as the name itself suggests, is an annual publication of the Central Bank of Nigeria and records data for the year ended on 31 December. It covers not only the financial sector of the economy but also the other sectors. It gives information on the fiscal and monetary policies adopted during the calendar year, domestic production and distribution, financial system, public finance including public debt at the federal level, balance of payments, international economic developments, organisational and administrative developments, and, finally, its own balance sheet for the year. It provides thus the general trends in the economy during the calendar year. The financial system data recorded in it relate to money supply and its determinants: financial system credit, money market, interest rates, central banking, commercial banking and other financial institutions.

The money supply and its determinants are discussed and recorded under the heads: credit to domestic economy, credit to the government sector, foreign assets, other assets, total money assets, quasi-money, money supply, currency outside banks, private demand deposits and total monetary liabilities. The institutional sector data cover saving and time deposits at commercial banks, the National Provident Fund, the Federal Savings Bank, the Nigerian Building Society, time deposits with merchant banks, premium bonds, saving certificates and saving stamps. The data on the components of the money market relate to the government short-term debt instruments: Bankers' Unit Fund (BUF), and Certificates of Deposits (CDs). The capital market transactions relate to the stock exchange transactions in Lagos, and the data on the Federal Republic of Nigeria Development Stock.

The data on interest rates structure include minimum discount rate, treasury bills rate, treasury certificates with one and two-year maturity, produce bill rate, lending rates (minimum and maximum), deposit rates, (minimum and maximum), and Federal Savings Bank rate. The items central banking, commercial banking and other financial institutions cover the balance sheet data of these institutions. The *Annual Report* thus covers a wide range of economic and financial data.

The second publication of the Central Bank of Nigeria, *Economic and Financial Review*, is published periodically, generally twice a year in December and in March or June. This publication records data on central banking, commercial banking, currency in circulation, external assets, international trade, money and capital markets, money supply, national saving, production in agriculture, electricity and minerals, consumer prices, public debt and public finance. Although the topics of the financial statistics covered in the previous and this publication overlap, except for the rate of interest, which is not recorded in the *Review*, its coverage of financial data is wider than that of the *Annual Report*. In commercial banking, it covers the activities of the monthly clearing of all the clearing houses in the country. This document covers the institutional sector savings only partly and records savings data of commercial banks and the Federal Savings Bank. The *Annual Report* gives the savings data of the other financial institutions also. However, the structure of rates of interest is missing from the *Review*.

The third publication of the Central Bank of Nigeria, *Monthly Report*, is published monthly and covers summary of developments (in a given month), money and banking, prices, flow of foreign exchange and external reserves and developments in the world markets of Nigeria's major agricultural exports of commodities. It, being a monthly publication, gives less coverage of the economy as compared with the previous two publications.

The *Development in the Nigerian Economy in the First Half of the Year* is an annual publication of the Central Bank of Nigeria and covers the same topics as the *Annual Report*.

The other relevant sources of financial statistics in Nigeria are the *Annual Digest of Statistics*, published by the Federal Office of Statistics, and the reports of the financial institutions, published monthly and annually.

After the detailed analysis of the sources of financial statistics in Nigeria, one can quickly summarise the sources of financial data of other African countries. The details of financial data are almost similar in the central bank publications of the other African countries. In the case of Algeria, the standard source of statistics is the *Bulletin of General Statistics*, published by the Department of Statistics. The *Information, Notes and Statistics*, published by the Central Bank of the West African Monetary Union, is the standard source of financial data of Benin Republic.

For Burundi, such data can be obtained from the *Quarterly Bulletin*, published by the Central Bank, and *Studies and Statistics*, published by the Central Bank of the Central African States. The latter publication provides data on Chad, Congo and Gabon also. The Egyptian financial data can be obtained from the *Economic Review*, published by the Central Bank. The *Quarterly Bulletins* of Ethiopia and Gambia, published by the central banks of these countries, give financial data. The *Economic and Financial Review*, published by the Central Bank of Kenya, gives the data on national accounts, apart from covering the routine data. The other source of financial statistics in Kenya is the *Statistical Digest*, published by the Bureau of Statistics.

In Liberia, the standard sources are the *Statistical Bulletin*, published by the National Bank of Liberia, and the *Quarterly Statistical Bulletin*, published by the Ministry of Planning and Economic Affairs. The Ghanaian financial statistics can be obtained from the *Annual Report*, produced by the Bank of Ghana. The same institution also publishes the *Quarterly Economic Bulletin*. The other source of financial data in Ghana is the *Digest of Statistics* produced by the Central Bureau of Statistics. Similarly, for Sudan, the sources of financial data are the *Annual Report*, *Economic and Financial Review* and *Statistical Digest*. The *Annual Report*, *Quarterly Bulletin* and *Quarterly Economic and Statistical Bulletin* form the standard financial sources in Uganda.

Similarly, the central banks of other African countries produce annual, quarterly and monthly reports, which contain financial data. They differ in coverage of the financial and non-financial statistics, but the crucial data on finance are available for every country of Africa. The financial institutions, apart from central banks, also publish data relating to their participation in economic and financial activities in these countries.

International Sources

Apart from the individual country publications, financial data are published at the international level also. There are two useful international financial sources: the *Government Financial Statistics Yearbook*, and *International Financial Statistics*.

The purpose of the *Government Financial Statistics Yearbook*, prepared by the Government Financial Statistics Division of the IMF's Bureau of Statistics, is to provide sound and comparable statistics on the finances of member governments and to lay the foundation for improved compilation of such data. The *Yearbook* presents the beginnings of a data base on government operation, information on the units, accounts and data on governments and on their non-financial public enterprises, public financial institutions and derivation of tables showing the adjustment made in data from national sources to compile the statistics in the format presented in the *Yearbook*. Detailed data are presented only for central governments, and summary data are given for state and local governments. Data for central governments are, however, incomplete in coverage in a number of cases. With time, it is hoped to expand the data to cover a greater portion of central government units and more state and local governments both in summary and in detail. The material for each country is organised in three parts: statistical tables, institutional tables and, where information is available, a derivation table followed by a statement on the coverage of the central government statistics. Information contained in the *Yearbook* has been provided by government financial-statistics correspondents in each government or central bank and has been discussed with IMF staff members at regional seminars on visits or through correspondence. Data are based primarily upon a manual on government financial statistics.

International Financial Statistics is another basic source. It is published monthly by the Bureau of Statistics of the IMF. It starts with the international data and then gives data relating to the individual country. The international statistics recorded in this publication relate to par values and central rates, participants' special-drawing rights (SDR) positions, members' Fund positions, transactions of all the member countries by components as drawings, repayments by purchases, net drawings, currencies drawn, repurchases by currency of repurchase and net Fund sales. It also records data of standby arrangements, Fund borrowing and world gold-holdings.

In addition, it gives the data of international reserves in detail thus giving the total reserves, gold, special drawing rights, reserve positions in the Fund, foreign exchange, and use of Fund credit. It also records the interest rates, money market and Eurodollar rates, central-government bond yields, end-of-month forward rates and daily exchange rates. These are published only for some selected countries. The publication also records the prices of major world-trade commodities, import and export price indices, changes in money supply, changes in consumer prices (percentage change in twelve months) and world trade. The last item records f.o.b. values of exports and of values of imports for all the member countries.

The *International Financial Statistics* furnishes information on individual countries, collected from the central banks of these countries and other institutional publications. The publication not only lists the sources of data; it also carries brief notes on the statistics of individual countries. Its coverage of the data includes, in the case of individual countries, exchange rate, trade conversion factor, international liquidity, monetary authorities, commercial banks, monetary survey, other financial institutions, prices and production, international transactions and balance of payments, government finance, national accounts and population. The details of these transactions are obtained from the countries concerned and thus differ for various countries depending on the availability of information with the IMF. For instance, for Kenya, all of the above information is recorded, while for Mauritania, data on other financial institutions, prices and production, and national accounts are missing from the publication. For some countries, the data are published in the form of notes: for example, Benin Republic and Senegal in West Africa.

The other international data sources are the statements published by the IBRD (World Bank) and the International Development Association (IDA). The data, however, are limited to the loans and credit provided for the calendar year. The IBRD publishes a *Statement of Loans* granted by it, and the IDA publishes a *Statement of Development Credit* granted to various nations.

The records of financial data of African countries, as is clear from the domestic and international sources, appear to be impressive. The financial data are up to date and cover most of the activities involving the use of money and finance in these countries. One can easily cross-check the financial data, as there are more different sources of data. The data on finance are therefore more reliable than other data in the African countries discussed in the different chapters of the present work. Nevertheless, they leave some gaps to be filled up and require serious efforts in some directions. These are taken up later.

COLLECTING THE DATA

The central statistics office in most African countries has overall responsibility for official financial statistics, as for all other official national statistics. The central banks are the main collectors of financial statistics and the sole collector of such statistics from banks; but several government departments too collect statistics from a variety of other financial institutions. As a collector of financial statistics, the central banks are in a very strong position. They are substantial producers of the raw data, because they are banks and have customers – principally, of course, other central banks and the government. They are also one of the main users of the whole range of financial data, either in their operational role in relation to economic management or in their role as supervisor of the banking system. This triple role of producer, collector and user is a great asset to central banks, but has its own problems. Certainly, it colours much of what they actually do as collectors; it may lead them, at one end, to overestimate the ability of every bank to complete the statistical returns that they themselves handle easily enough and, at the other end, to under-rate the special needs of other users whose version of how the monetary system works may be based on other theories of economic management. But to have both experience of compiling the returns and strong views about the correct use of the resulting statistics is no bad thing for a collector. There are frequent clashes of interest between producer and user, often but not always arising from the costs of compiling or the confidential nature of the raw material. The banks are thus conditioned to deal understandingly with these problems as they are seen by the institutions from whom they collect statistics; and possibly they have a little more confidence in the banks, because they know that they are completing the same returns themselves. But in the end the banks have to spend, as most other collectors of statistics do, as much time putting the producers' case to the users as putting the users' case to the producers. The problem areas are, of course, questions of definition, detailed classification, frequency of reporting, promptness, accuracy of return and confidentiality of detail.

BANKING STATISTICS AND USES

Let us now turn to banking statistics as the branch of financial statistics with which central banks are mostly involved. The volume of the range of statistics has grown as regards both the number of banking institutions covered, the frequency and the complexity of the statistical returns.

The characteristics of banking figures are rather special. The returns made to central banks by reporting banks are very largely concerned with various aspects of their financial assets and liabilities, which are, of course, their stock in trade and about which they themselves need to be well informed in order to run their business efficiently. Generally speaking, banking figures are both prompt and accurate. The fact that they are derived from, or related to, balance sheets is a help in this direction, although the detailed classifications that they contain provide plenty of scope for error.

The returns are used for a variety of purposes, some administrative and some statistical, and it has always been the policy to have all-purpose returns rather than separate returns designed for separate purposes. This represents an economy both to reporting banks and to the central banks themselves.

The individual returns themselves mostly provide the information that is needed for administrative purposes; for example, the returns show the banks' observance of reserve ratio requirements, or the disposition of their balance sheets in relation to bank supervision criteria. Aggregates compiled from the individual returns from all banks provide two quite different kinds of information.

First, they bring sharply into focus the operations of the banking system itself. They provide a sequence of snapshots of the condition of the system, showing, for example, the liquidity of its assets in relation to its liabilities; but they do not directly show transactions. This is an important distinction. Such aggregates are used by the authorities, in conjunction with the individual returns, for particular purposes, such as credit control and bank supervision. They are also seen by the contributing banks as the background to their own operations, enabling them to measure their market shares in the various markets, or subdivisions of markets, covered by the returns.

Second, because the banking system is an important repository of funds for most categories of residents, and for overseas residents also, the aggregates for the system, suitably analysed by categories, contribute a unique body of information about the economy at large. In conjunction with the central bank's own figures, the aggregated returns from reporting banks provide the sector analyses of deposits and bank lending used in the flow-of-funds accounts. This is the use that links banking data to other financial and economic information in a regular and systematic series of quarterly national accounts. They also provide the figures of money stock, which are nowadays in the forefront of decisions on monetary policy. It is important to stress that figures that are important for current assessments are generally needed more frequently than quarterly, and most of the banking statistics are compiled both quarterly and monthly.

Other Uses of Financial Statistics

We have mentioned that the central banks are not only collectors of the financial statistics; both the government and the central banks are also the major users of financial statistics:

(1) for monitoring the economy and changes in the financial system;
(2) for assessments leading to policy decisions;
(3) for administrative purposes in connection with credit control and exchange control;
(4) for supervision of financial institutions; and
(5) for research (although we should be unlikely to collect financial data solely for research, it is an important additional use for data collected for other purposes).

Other users generally need financial statistics for assessment or research, or for commercial use in a particular line of business; and one must mention especially the needs of financial institutions themselves, in running their own affairs, for the aggregated data to which they contribute. For many users, such as financial commentators and analysts, the statistics are only the grist to the mills of analysis; and for government and banks, the problems of collection run straight through without a break into the problems of analysis. To meet these needs, there are broadly two kinds of financial information: prompt and fairly frequent reports on what has been happening in the recent past; and a wider range of data structured to indicate the characteristics and patterns of business of the financial system itself. We explain in detail some of these uses in African countries.

The financial statistics serve as an indicator of the national and sectoral financial positions. The health of a nation's economy can be analysed by assessing the foreign exchange reserves. If the external reserves of any country are falling, the economy may be labelled 'sick'; if, on the other hand, these reserves are rising, the economy may be thought of as 'healthy'. For instance, in Nigeria, external transactions in foreign exchange during August 1977 resulted in a net outflow of 101·00 million Naira, compared with the net outflows of 125 million Naira and 18·9 million Naira in the preceding month and the corresponding month of 1976 respectively (Nigeria, 1977). The trend is also affirmed by the yearly data. The continuous fall in foreign reserves has created an alarming situation in the country. Similarly, we can analyse the assets and liability position of different sectors to assess their progress.

Apart from analysing the financial health of a country or sector, the financial statistics serve as an important indicator of the sources and uses of funds of the country and of different sectors in an economy. Thus, they show the important sources of incurring liabilities and the relative importance of each source over time. By analysing the sources of finance of other companies, a new establishment can design its borrowing policy concerning the availability of funds.

The role of financial intermediaries can also be assessed by analysing their sources of finance. For this purpose, sources of finance may be divided into internal sources and external sources. Internal sources relate to own saving and the provision for depreciation. External sources refer to borrowing from the other sectors, which may be subdivided into direct and indirect sectors. If the surplus sector acquires claims on debtors through financial intermediaries, it is termed as indirect. On the other hand, if the surplus units acquire claims on debtors that are not financial intermediaries, it is known as direct. The role of financial intermediaries can be assessed by taking the proportion of indirect finance to total finance or to total investment. The financial statistics are also required to identify the sector origin of the expansionary and contractionary forces in an economy. For instance, if the enterprise sector invests more than its capacity to finance internally, it has to resort to the method of external finance. If it is unable to tap resources from the non-financial sector in amounts sufficient to cope with its requirements, the sector will fill up the gap either by decreasing its holdings of money or quasi-money or by borrowing from the financial sector by way of loans and sales of claim to it. This excess investment over savings in the enterprise sector augments income of households, which, if not saved, leads to demand for commodities and, if this goes unmatched by the increase in output, leads to inflationary conditions.

The financial statistics not only help identify the sector that contributes to inflationary trend in the economy, but also reveal it at the national level by presenting the data on the money supply. A substantial increase in the money supply in African countries, where the productive capacity is limited, brings about inflation. The financial statistics also help in designing the policies to curb inflation and assess their efficacy. In the case of inflation, the central bank of a country may resort to open market operation, bank rate policy and cash reserve ratio (quantitative instruments at its disposal) and moral situation (the qualitative method). The efficacy of these measures can be judged by whether these reduce the supply of money in the economy.

The use of financial statistics in financial planning can hardly be overemphasised. For financial planning at the sectoral level and at the national level, we require a financial matrix, which can be constructed with the help of the financial statistics already available. The financial matrix was used in the formulation of the Fifth Five-Year Plan of India. But in African countries, where rigorous planning techniques are still to be used in the formulation of the plans, the financial matrix is yet to be constructed.

Financial statistics may also be used for international comparisons of the various components. Thus, one can compare the liquidity ratio of different countries or assess the relative position of commercial banks' loans and advances. For this purpose, the IMF publishes *International Financial Statistics*.

LIMITATIONS

There are many gaps and inadequacies in the range of financial statistics collected and published. The financial data collected by African countries relate to the organised financial sector only. However, in the developing African countries, where a majority of the population lives in villages, and where the modern banking system is either unheard of or is newly introduced, people depend for finance on the native money-lenders. These traditional money-lenders form the unorganised markets in these countries. Their activities are not only restricted to providing finance to the needy persons; they themselves own land and earn income from agriculture and trading. The income that they earn is a mixture of interest income, trading and agricultural income. Moreover, their *modus operandi* of lending money also differs from that of the organised financial institutions. They generally offer loans against mortgage of land and other property. To exclude the activities of these financial (unorganised) intermediaries is to ignore the major source of finance in these countries.

As regards the data on the rate of interest, the published data again cover the structure of the rate of interest of the organised financial institutions only. The rate of interest charged by the unorganised sector, which is not covered by these publications, differs from one individual borrower to another, depending upon the nature of the security that he is offering and the risk involved. Generally, the rate of interest charged by the traditional money-lenders is much higher than that charged on similar borrowing by the organised financial institutions. Sometimes, these money-lenders may even borrow from banks and relend the money at a higher rate of interest. Similarly, the data on gold-holdings in the country refer to the holdings by the banking system. In the countries, like Ghana, where people also have access to gold, the data of the published sources may present underestimates of the gold-holdings in the economy. In villages, people do not know of many ways of utilising their savings and may generally invest in land or gold, which provide them with maximum security. On the contrary, in urban areas, people know of various means of utilising their savings; they may buy shares of varying maturities and deposit money in the bank. In the villages, however, if an individual does not want to invest in land or gold (where he has access), he may simply dig a pit and hide his money there. The institutional publications do not cover this aspect of hoarding also.

Moreover, the official publications have no information on the illegal money supply in the country. The money supply refers only to the banking system, but newspaper reports sometimes reveal the existence of illegal printing presses in these countries, which are also a source of pumping money into the economy. The same can be said of the foreign exchange reserves in the country. Smugglers often cause serious problems in African countries. But all the facts often cannot be converted into figures. We may be willing to record all the statistics on the above aspects, but the practical problems hamper us even in identifying these transactions, far from knowing their exact or approximate magnitude.

CONCLUDING REMARKS

It is heartening to note that there are various sources – domestic and international – that record financial data on African countries. One can easily cross-check the data published in the various sources and assess the reliability, which is fairly high. However, there is a strong case for increasing the coverage of the financial statistics to include also the activities of the unorganised money markets. But practical difficulties in knowing the magnitude of their transactions hamper such attempts.

REFERENCES AND FURTHER READING

Arya, P. L. (1959), 'Integration of national accounts with flow-of-funds accounts', *Business Analyst* (May – August).

Arya, P. L. (1976), *Social Accounting for Developing Countries* (New Delhi: Macmillan Co. of India).

Gurley and Shaw (1955), 'Financial aspects of economic development', *American Economic Review*, pp. 515 – 38.

India (1959), *Report of the Working Group on Flow of Funds* (Delhi: Central Statistical Organisation of India).

International Monetary Fund (IMF) (annual), *International Financial Statistics* (Washington, DC: IMF Bureau of Statistics).

Nigeria (1975), *Annual Report and Statement of Accounts for the Year Ended 31st December* (Lagos: Central Bank of Nigeria).

Nigeria (1977), *Monthly Report* (Lagos: Central Bank of Nigeria, October).

Nigeria (annual), *Economic and Financial Review* (Lagos: Central Bank of Nigeria).

Nigeria (monthly), *Monthly Report* (Lagos: Central Bank of Nigeria).

Thornton, M. (1977), 'Financial statistics: collection and data', *The Statistician*, vol. 26, no. 4, pp. 243 – 57 (London: Institute of Statisticians).

MAJOR SOURCES OF BANKING AND FINANCIAL STATISTICS AND PUBLICATIONS ON BANKING STATISTICS

ALGERIA
Department of Statistics, *Bulletin of General Statistics*.

BENIN
Central Bank of the West African Monetary Union, *Information, Notes and Statistics*.

BOTSWANA
Botswana National Development Bank, *Annual Report*.

BURUNDI
Banque de la République du Burundi, *Bulletin trimestriel*.
Central Bank of the Central African States, *Studies and Statistics*.

CHAD
Banque de Développement du Tchad, *Rapport d'activité*.
Central Bank of the Central African States, *Studies and Statistics*.

CONGO
Central Bank of the Central African States, *Studies and Statistics*.

EGYPT
Central Bank of Egypt, *Report of the Board of Directors*.
Central Bank of Egypt, *Economic Review*.

ETHIOPIA
Commercial Bank of Ethiopia, *Market Report* (monthly).
Addis Ababa Bank, *Annual Report*.
Commercial Bank of Ethiopia, *Annual Report*.
Ethiopian Investment Corporation, *Report of the Directors*.
Mortgage Company of Ethiopia, *Annual Report*.
National Bank of Ethiopia, *Quarterly Bulletin*.

GABON
Central Bank of the Central African States, *Studies and Statistics*.

GAMBIA

Currency Board, *Report* (annual).

Post Office Savings Bank, *Report*.

Government Savings Bank, *Report for the Year* (annual).

Central Bank of the Gambia, *Quarterly Bulletin*.

GHANA

Post Office Savings Bank, *Annual Report*.

Bank of Ghana, *Quarterly Economic Bulletin*.

Bank of Ghana, *Report of the Board for the Financial Year* (annual).

Bank of Ghana, *Annual Report*.

Ghana Commercial Bank, *Commerbank News* (bi-monthly).

Ghana Commercial Bank, *Monthly Economic Bulletin*.

National Investment Bank, *Report of the Directors* (annual).

Ministry of Finance, *Annual Report: Insurance in Ghana*.

Central Bureau of Statistics, *Digest of Statistics*.

KENYA

Central Bank of Kenya, *Annual Report*.

Central Bank of Kenya, *Economic and Financial Review*.

Bureau of Statistics, *Statistical Digest*.

LIBERIA

National Bank of Liberia, *Statistical Bulletin*.

Ministry of Planning and Economic Affairs, *Quarterly Statistical Bulletin*.

LIBYA

Bank of Libya, *Balance of Payments: Economic Research* (annual).

Bank of Libya, *Economic Bulletin: Economic Research* (bi-monthly).

National Bank of Libya, *12th Annual Report of the Board of Directors*.

MALAWI

Reserve Bank of Malawi, *Annual Report and Statement of Accounts*.

MAURITIUS

Bank of Mauritius, *Quarterly Review*.

Development Bank of Mauritius, *Report and Accounts*.

NIGER

Banque de Développement de la République du Niger, *Rapport annuel exercise*.

Banque de Développement de la République du Niger, *Informations économiques: Bulletin d'études trimestriel*.

NIGERIA

Central Bank of Nigeria, *Monthly Report*.

Central Bank of Nigeria, *Annual Report and the Statement of Accounts*.

Central Bank of Nigeria, *Commercial Bank Activities* (monthly).

Central Bank of Nigeria, *Economic and Financial Review* (bi-annual).

Central Bank of Nigeria, *Development in the Nigerian Economy in the First Half of the Year* (annual).

Federal Office of Statistics, *Annual Digest of Statistics*.

SIERRA LEONE

Bank of Sierra Leone, *Annual Report and Statement of Accounts*.

National Development Bank Ltd, *Report and Accounts*.

Bank of Sierra Leone, *Economic Review: Quarterly Balance of Payments of Sierra Leone* (annual).

Economic Trends (bi-monthly).

Annual Report.

Sierra Leone in Figures (annual).

Charts on the Economy of Sierra Leone.

SOMALIA

Somalia National Bank, *Report and Accounts*.

Somalia National Bank, *Report and Balance Sheets*.

SUDAN

Bank of Sudan, *Annual Report*.

Department of Statistics, *Economic and Financial Review*.

Department of Statistics, *Statistical Digest*.

SWAZILAND

Swaziland Credit and Savings Bank, *Annual Report and Accounts*.

TANZANIA

Bank of Tanzania, *Economic and Operations Report.*

National Bank of Commerce of Tanzania, *Report.*

UGANDA

Uganda Commercial Bank, *Annual Report.*

Bank of Uganda, *Annual Report.*

Bank of Uganda, *Quarterly Bulletin.*

Statistics Division, Ministry of Planning and Economic Development, *Quarterly Economic and Statistical Bulletin.*

ZAIRE

Banque Commerciale du Zaire, *Rapport et bilan.*

Banque de Kinshasa, *Exercise.*

ZAMBIA

Bank of Zambia, *Quarterly Statistical Review.*

Bank of Zambia, *Report and Statement of Accounts* (annual).

Commercial Bank of Zambia Ltd, *Annual Report.*

National Commercial Bank Ltd, *Report and Accounts* (annual).

INTERNATIONAL

International Bank for Reconstruction and Development, *Statement of Loans.*

International Development Association, *Statement of Development Credit.*

International Monetary Fund, *Government Financial Statistics Yearbook* (annual).

International Monetary Fund, *International Financial Statistics* (annual).

CHAPTER 11

External Trade Statistics

African countries perceive the problem of economic and social development as one of the challenges facing them. Having achieved their political freedom, they now turn their attention to the international economy, a structure that continues to reflect the trading patterns of an earlier colonial era. This has brought into sharp focus the need for assembling reasonably-accurate external-trade statistics, which can guide policy-makers and planners.

Meanwhile, the expansion of multinational companies has rendered conventional trade figures partially obsolete. A growing proportion of transactions that cross boundaries is internal to the company concerned. Moreover, the political power and the economic weight of these multinational corporations are giving rise to growing concern. Detailed information is therefore needed on these corporations, including their branches and affiliations in various countries, and on their transactions.

This chapter attempts to discuss the nature and scope of external trade statistics in Africa, their sources, uses and limitations, as well as gaps in the existing data collection of such statistics and methods of improving them.

NATURE AND SCOPE OF EXTERNAL TRADE STATISTICS

External trade statistics give information on the volume of goods and services traded, their composition, their direction and the balance of payments. The volume of trade refers to the total exports and imports of goods and services during a particular period of time; the composition of trade refers to the various goods comprising exports and imports; and the direction refers to the trading partners.

The balance of trade is regarded as the difference between the export and import of merchandise (commodities). The balance of payments consists of current account and capital account. The balance of payment on current account can be obtained by adding the balance-of-services account (invisibles) to the balance of trade. Thus, the balance of payment on current account reflects the total value of import and export of goods and services and their net amount. Any surplus or deficit of the current account is matched by the outflow or inflow of foreign capital in the capital account. The capital account of the balance of payments deals essentially with the inflow and outflow, or the borrowing and lending, of the capital.

Information on prices of the commodities traded, indices of export and import, foreign exchange rate and the terms of trade are some of the related data that fall under the periphery of the external trade statistics.

SOURCES AND METHODS

The records of customs and excise form the source of primary data, and the publications of the bureaus of statistics and the central banks of African countries form the secondary sources of data. Apart from these domestic sources of statistics, data are also available at the international level in the following publications of the International Monetary Fund (IMF), the International Bank for Reconstruction and Development (World Bank, IBRD) and the United Nations (UN):

Direction of Trade (annual, IBRD and IMF)
International Financial Statistics (IMF)
Survey of African Economies (IMF)
Monthly Bulletin of Statistics (UN)

These international publications cull data from various countries and then present them in a manner that is useful for international comparisons. Much therefore depends on the individual country data that is supplied by the bureau of statistics of different countries.

It is difficult to examine all the sources of foreign trade statistics of African countries, for they are numerous, and the nature of statistics collected by them is more or less similar. The data also suffer from common weaknesses. It will suffice to take the example of one country and examine the sources and methods of statistics in detail. The choice of the country rests on our access to the basic sources of data, and hence Nigeria is the obvious choice.

The accuracy and reliability of any statistics depend largely on the primary sources of information. Except for the world commodity prices, which are available for a number of Nigeria's primary commodities in several international publications, the results of records of external trade transactions of the Board of Customs and Excise form the primary source of external trade statistics of Nigeria. The documents are processed and published by the Federal Office of Statistics, and the most detailed publication on this is the *Nigerian Trade Summary*. It is published on a quarterly basis. Monthly data are, however, published for a limited number of commodities in the *Digest of Statistics* and the *Review of External Trade*. Further data are contained in the *Supplement to the Digest of Statistics*.

The Nigerian External Trade Classification, to which the trade summary is patterned, is drawn up in line with the Standard International Trade Classification (SITC). All commodities are classified into ten sections (i.e. 0 – 8 according to kind, and those that cannot be identified form section 9). Since commodities in a section are similar in kind, such section is divided further into more homogeneous groups according to degrees of similarity; each group is further subdivided into commodities.

In Ghana, for example, the examination of the details of the various annual import programmes shows that the implementation and monitoring of the import trade are the direct responsibility of the Ministries of Trade and Industries and the Department of Customs and Excise. The programmes are set out according to the SITC. Only in 1966 was the foreign exchange budget expressed in SITC categories, and then only at the single digit level. In all other years, the programmes have been chiefly specified according to a curious mixture of end-use, industry-of-use and specific-commodity categories. Before they can be given operational content, the programmes have to be converted into SITC classes. One consequence of this practice is that it was impossible to mention the programmes drawn up, for the trade statistics follow the SITC, and the end-use composition of imports is only estimated annually after long delays. The same can be said of many other African countries, which operate similar procedures.

The role of customs and excise services in the compilation and processing of trade statistics is to ensure that the external trade documents are completed and duly filled by importers and exporters or their agents for purposes of customs clearance. These documents are then transmitted to the Office of Statistics for scrutiny, processing, tabulation and publication. These data are thus a byproduct of customs and excise functions; but as goods arriving at or leaving Nigeria's territory are subject to customs clearance (apart from a few exceptions like ships, aircraft, diplomatic and transit goods), the customs records constitute virtually the complete information on the quantity and value of the legal external-trade transactions. They do not, of course, include data on trade through smuggling, for obvious reasons.

The exports and imports are recorded at the time the goods cross the frontiers. However, there are several defects in these data for calculation of the unit value of each commodity. Customs records and, of course, the *Nigerian Trade Summary*, do not furnish specific and separate price data for each individual product comparable through time, but rather aggregate values through quantities associated with a given consignment or a commodity defined under the trade classification. Besides, several entries have only unit data available, and the lack of quantity data makes it impossible to derive unit values directly. In several cases the commodity classification is so broad that it does not in actual fact refer to any product; an average unit price in such cases would be meaningless.

Since the above-mentioned publications of the Federal Office of Statistics draw on the records of customs and excise, which we have already discussed at length above, all the weaknesses of the customs and excise data are carried to the data issued by the Federal Office of Statistics.

The data on the balance of trade and the balance of payments of Nigeria are compiled and published by the Central Bank of Nigeria in the following four publications: *Economic and Financial Review*, *Annual Report and the Statement of Accounts*, *Monthly Report* and *Developments in the Nigerian Economy in the First Half of the Year*.

The *Economic and Financial Review*, in its section on the international trade, gives information on the balance of payments, foreign trade, imports by SITC, exports of major commodities, exports by regional grouping and imports by regional grouping. Its foreign trade figures are also based on the data supplied by the Federal Office of Statistics. However, the additional information on the other components of the balance of payments is compiled by the Central Bank itself. The information essentially relates to the flow of foreign exchange and external debt position.[1]

The *Annual Report and the Statement of Accounts* gives the balance of payments, merchandise trade in respect of exports and imports, direction of trade in respect of non-oil products and crude oil products, services account, direction of trade unrequited transfers, current account, capital account and external assets.

Developments in the Nigerian Economy in the First Half of the Year covers the same items as the *Annual Report*, except direction of trade and oil and non-oil bifurcation of imports and exports.

The Central Bank's *Monthly Report* gives data on Nigeria's external trade, flow of foreign exchange through the Central Bank, monthly value of major export commodities, monthly exports of major items in quantities, and the value of major import groups by SITC sections. Prior to 1964, the Federal Office of Statistics provided foreign trade indicators such as unit value, quantum and terms-of-trade indices, but these have since been discontinued. The information on these items is, however, available for some other African countries.

Analogous to the Central Bank of Nigeria, the balance-of-payments statistics are compiled by the central banks of most of the other African countries. The periodicity and the details of the items recorded differ in their publications, depending on whether a publication is monthly, quarterly or annually.

The central banks and the bureaus of statistics of these countries provide data of exports and imports to the international institutions like the UN and the IMF, which in their turn publish them on an international basis. Since a common format is used for recording data by the international agencies, the external trade statistics recorded by them is very useful for intercountry comparisons. For instance, the *International Financial Statistics* publishes data on foreign transactions, among other data, for African and other countries. It gives in detail the exports of important commodities, imports (c.i.f. and f.o.b.), export index by principal commodities, index of export prices and the balance of payments.

USES OF EXTERNAL TRADE STATISTICS

The foreign trade statistics serve as a useful guide to the monetary and fiscal policies of the country. If the value of exports shows a declining trend, and the value of the goods imported shows a continuous upward tendency, the situation calls for remedial measures both on the monetary side and on the fiscal side. One may recommend the devaluation of the currency (to cheapen exports in the world market and to make the imports dearer in the domestic market), to put up import duties in order to increase the price of imported goods in the domestic market, or to ban imports of non-essential products. Each of the policies adopted has widespread consequences, and their efficacy would depend on the elasticity of demand for imports and exports in the domestic market and in foreign countries, the response of domestic production as well as the retaliation from the other trading partners. The discussion on the policies to be adopted and their consequences is outside the scope of this chapter.

The statistics on external trade also give in detail the transactions carried out with different countries, and one can judge the relative importance of a particular trading partner in one's country's economy. If a country depends heavily on one or two countries for its imports and exports, then that foreign country may exercise considerable political influence on the domestic market of the country. It is therefore advisable for the domestic market or home country to find other trading partners, to improve the terms of trade, and to reduce the political influence of that country.

Similarly, the analysis of the composition of trade is useful in judging whether a country is depending too much on the export of one commodity. Nigeria is a case in question. Out of the total exports of 4922·5 million Naira in 1975, the oil sector exports constituted 4560·9 million Naira (about 92·5 per cent). A recent fall in the demand has resulted in some decline in the prices and reduction in the production of crude oil, which in their turn have reduced the foreign exchange earnings. Such a situation could have been avoided if Nigeria had paid attention to the development and export of other products.

Reliable trade data are also useful for building econometric models needed for forecasting. These forecasts may help in formulating five-year plans for economic development. These are only a few of the innumerable uses of the foreign trade statistics.

INCONSISTENCIES IN AND LIMITATIONS TO THE USE OF THE DATA

Discrepancies in external trade data were noticed nearly sixty years ago (Zuckermann, 1920). This was followed by a series of studies by the League of Nations (1935 – 8) where inconsistencies in particular trade flows were revealed and examined and a catalogue of the suspected reasons was provided. More recently Morgenstern (1963) examined the accuracy of trade statistics and noted: 'Further investigations of this nature must be undertaken to decide whether foreign trade statistics can be trusted in proving fine theoretical points and formulating policy.'

In 1970 a United Nations reconciliation study reached the following conclusions:

> In many countries, few taxes or quantitative controls are applied to exports, with the result, that the interest of customs authorities in the control and documentation of exports is limited . . . . It may therefore happen more frequently than is normally suspected that a significant number of export transactions are neither documented nor recorded . . . Moreover, export documentation may suffer from inadequate commodity descriptions and other tabulated information in so far as it is not subject to thorough checking procedures administered by customs authorities. (United Nations, 1974, p. 7)

Parniczky (1980) in a study on inconsistency of world trade statistics noted:

> Perfect consistency is an ideal objective, which can never be achieved in practice. First, there is a time-lag between the declaration of an export transaction and the observation of the corresponding import. As a result, a specific period of time has different coverage in terms of exports and imports. However, this source of inconsistency has special importance for pairs of remote

countries only. Moreover, reconciliation of inconsistent records can be achieved by means of a suitable logging of imports behind export data, for any given trade flow.

The role of customs administration, as a potential source of inconsistency, on the other hand, merits closer attention. As a rule, foreign trade statistics are generated by the customs authorities, based on the declarations filed by the exporters and importers for the primary purpose of administrative control and tariff revenue collection. There are notable exceptions: in Hungary, for example, statistical data are compiled directly from the trading establishments and the role of customs administration is confined to certifying the movement of merchandise across the border.

With respect to disaggregated trade flows, it should be noted that the level of the commodity breakdown has a certain impact on the size of inconsistency. The finer the classification used, the larger the discrepancy that can be expected [see United Nations, 1974, Table 5, p. 11]. Although both sides may use the same standard classification, perfectly matching interpretation of commodity classes has not been achieved and the probability of different understanding grows with the detail. What is correct commodity identification for one country may be a 'misclassification' for the partner.

The sources of inconsistency identified so far affect both quantity and value data. Values, in addition, are affected by the valuation conventions mentioned earlier. The discrepancy between export and import data is due to the treatment of the transportation and insurance costs. Whereas all countries report exports on an f.o.b. basis, imports are normally valued c.i.f.[2] Consequently the value of exports and imports could not match even under ideal conditions.

Although the reasons discussed above should not be underestimated, they alone cannot generate the large gaps between export and import data [described in the previous section]. The principal source of inconsistency appears to be the role of entrepôt trade (middleman trade) in commercial transactions. The operations of large enterprises in 'free zones', customs bonded store-houses and bonded processing establishments may confuse the mutual identification of partner countries. Frequently the exporter is not aware of the final destination of the merchandise and the importer has a multiple choice in identifying the country of provenance, depending on the precise definition. In a broad sense, the 'mysterious disappearance of ships' [already discussed] can be explained in terms of entrepôt trade: the ocean is a gigantic free zone where ships are sent by the exporter, and they operate in that zone forever.

The expansion of the multinational companies in Africa and elsewhere has also rendered conventional trade statistics partially obsolete. A growing proportion of transactions that cross national boundaries are internal to the company concerned. Moreover, the political power and the economic weight of the multinational corporations or companies are giving rise to growing concern. For example, data on external trade can be affected by the underhand flow carried out by the multinational companies that have their own accounts and branches in foreign countries. Consequently, some commodities purchased by them for use in these countries are not covered by the permission of the central bank for foreign exchange. Similarly, any purchase done by its branch in a foreign country may not require foreign exchange, and hence permission from the central bank, as its counterpart may pay for its transactions in the local currency. The magnitude of these transactions increases with the increase in the foreign exchange restrictions. The nature of these weaknesses, arising from illegal transactions, of the foreign trade and the balance-of-payments statistics does not lend itself to identification, and one cannot say with certainty how much the official figures of exports and imports and inflow and outflow of foreign exchange are flouted by these transactions. Still, one can make a guess that the combined effect of these factors is well over 5 per cent of the total imports of goods and total outflow of foreign exchange.

Detailed information is needed on these corporations (including their branches and affiliates in various countries) and their transactions. Information about marketing channels of Africa's export commodities, including trade margins at the various stages, is also of vital concern to countries concerned. Similarly, African countries are highly interested in the marketing structure of their imports, a field of statistics that has virtually not been touched. Additional requests are raised by the pressure for indexation of commodity prices. For this and other purposes, unit values are rarely satisfactory measures. There is need for more precise identification of import and export prices. There are, however, many difficulties: the familiar problem of nominal prices declared for customs purposes, especially by the multinational companies, distorts the value of trade statistics for economic analysis. Moreover, it is not easy to evaluate the relative importance of multinational corporations' operations in the domestic market.

Much, however, has to be done to improve these statistics in Africa. One possibility is the matching of trade unit values as reported by both trade partners. Tanzania's experience in checking the wholesale prices of commodities exported and imported by mobilising all types of supplementary information is a good example of how the problem can be tackled (see Green, in press).

Nor is this all. Public interest in the trade figures is prompted by the hope that they give an indication of the country's current-account trade balance. But they are a most imperfect guide, because they measure the invoiced value of the goods flowing through the customs shed rather than the financial flows of the trade account. At times when exchange rates are moving and settlement periods shortening or lengthening, the differences are sizable. This is all the more important when the most publicised variable is the small difference between the two big figures for imports and exports. Finally, the external trade statistics are most inadequate on invisible trade.

CONCLUDING REMARKS

Although the data on external trade and balance of payments are very useful for planning and policy-making and there are several sources from which one can obtain these data, they suffer from common weaknesses, as these sources draw heavily on the records of the customs and excise for exports and imports and the central bank for inflow and outflow of foreign exchange.

However, we have to remember that time and again figures produced for one purpose have to be applied to another. For example, customs records, which are collected for a different purpose, are used also as external trade statistics, which serve another purpose. Therefore, data collected by customs officials, although used for the construction of external trade statistics, may not strictly conform to the requirements of statistics on external trade. Further, smuggling and other underhand business and payments carried out by multinational companies make the data on foreign trade unreliable. Nevertheless, compared to other data, and using the quality and reliability of other statistical data in African countries as criteria, one can conclude that the external trade statistics are fairly reliable.

NOTE: CHAPTER 11

1 The format of the balance of payments prepared by the Central Bank of Nigeria can be seen in the *Economic and Financial Review* (1977), pp. 74 – 5 of which give detailed layout.

2 For definitions and further information, see United Nations (1970), pp. 39 – 55. At present eighteen countries report imports f.o.b.; all other countries adopt the practice referred to above, or apply similar definition.

REFERENCES AND FURTHER READING

Ajani, H. A. and Asogu, J. O. (1977), 'External trade indices of Nigeria', *Economic and Financial Review*, pp. 5 – 22.

Ball, R. J. (1973), *The International Linkage of National Economic Models* (New York: North Holland).

Bureau of Census, United States Department of Commerce and Statistics Canada (1970), *The Reconciliation of US-Canada Trade Statistics* (New York).

International Bank for Reconstruction and Development (IBRD) and International Monetary Fund (IMF) (annual), *Direction of Trade* (Washington, DC).

IMF (annual), *International Financial Statistics* (Washington, DC).

IMF (1975), *Survey of African Economics*, vol. 6 (Washington, DC), pp. 359 – 68.

League of Nations (1935 – 8), *International Trade in Certain Raw Materials and Foodstuffs by Countries of Origin and Destination* (Geneva).

Leonard, W. R. (1953), International Comparison and Standardisation, in R. G. D. Allen and J. E. Ely (eds), *International Trade Statistics* (New York: Wiley).

Limmemann, H. (1966), *An Econometric Study of International Trade Flows* (New York: North Holland).

Morgenstern, O. (1963), *On the Accuracy of Economic Observations*, 2nd edn (Princeton, N.J.: Princeton University Press).

Nigeria (1976), *Developments in the Nigerian Economy in the First Half of the Year* (Lagos: Central Bank of Nigeria).

Nigeria (undated), 'Trade and monetary relations in the West African sub-region, 1966 – 1970', *Economic and Financial Review*, vol. 10, no. 1, pp. 5 – 22.

Nigeria (annual), *Annual Report and the Statement of Accounts* (Lagos: Central Bank of Nigeria).

Nigeria (annual), *Economic and Financial Review* (Lagos: Central Bank of Nigeria).

Nigeria (annual), *Digest of Statistics* (Lagos: Federal Office of Statistics).

Nigeria (annual), *Supplement to the Digest of Statistics* (Lagos: Federal Office of Statistics).

Nigeria (annual), *Nigerian Trade Summary* (Lagos: Federal Office of Statistics).

Nigeria (annual), *Review of External Trade* (Lagos: Federal Office of Statistics).

Nigeria (monthly), *Monthly Report*, (Lagos: Central Bank of Nigeria).

Parniczky, G. (1980), 'On the inconsistency of world trade statistics', *International Statistical Review*, vol. 48, no. 1, pp. 43 – 8.

Stern, J. J. (1972), 'Ghana's exports and terms of trade, 1955 – 70', *Economic Bulletin of Ghana*, no. 4.

United Nations (1949), *Relative Prices of Exports and Imports of Under-developed Countries* (New York).

United Nations (1970), *International Trade Statistics Concepts and Definitions*, Series M. no. 52 (New York).

United Nations (1974), *International Trade Reconciliation Study*, E/CN. 3/454 (New York).

United Nations (1976), *Yearbook of International Trade Statistics 1975* (New York).

United Nations (1977), *Concepts and Practices in International Trade Statistics*, ESA/STAT/AC. 5/2 (New York).

United Nations (monthly), *Monthly Bulletin of Statistics* (New York).

United Nations Conference on Trade and Development (UNCTAD) (undated), *Examination of Recent Developments and Long Term Trends in World Trade and Development*, Statistical Paper, General, TD/138/Suppl. 1.

UNCTAD (1965), *Handbook of International Trade and Development Statistics* (Geneva).

UNCTAD (1976), *Handbook of International Trade and Development Statistics* and *Supplement 1978* (Geneva).

Viner, J. (1952), *International Trade and Economic Development* (Glencoe, Ill.: Free Press).

MAJOR SOURCES OF EXTERNAL TRADE STATISTICS

ALGERIA

Secrétariat d'Etat au plan:

Indices du commerce extérieur de l'Algérie, no. 2 (1971).

Document statistique du commerce extérieur de l'Algérie (1970; annual).

Statistiques trimestrielles du commerce extérieur de l'Algérie: imports et exports: résultats paysproduits (1968, 1970).

Bulletin comparatif trimestriel du mouvement commercial et maritime de l'Algérie (1967).

Bulletin corporatif du mouvement commercial et maritime de l'Algérie (1967 – 8).

ANGOLA

Repartição de Estatística Geral, *Comércio externo* (1968, 1969, 1970; annual).

CHAD

Commerce extérieur contrôlé (1968, 1969; annual).

EGYPT

Central Agency for Public Mobilization and Statistics:

Monthly Bulletin of Foreign Trade (1967, 1968, 1969, 1970, 1971, 1972).

Egypt Foreign Trade (1970).

UAR Foreign Trade According to the Standard International Trade Classification (Revised) (1967, 1968, 1969; annual).

ETHIOPIA

Central Statistics Office:

External Trade Statistics (1967, 1968, 1969; monthly).

Annual External Trade Statistics (1967, 1968, 1969, 1970, 1971).

Monthly Trade Report (1971).

Port Office Assab, *Monthly Report: Import Commodities* (1968, 1969, 1972).

Department of Marine, *Trade of Port Massawa* (1967, 1969, 1970; bi-monthly).

GAMBIA

Customs Department, *Report* (1967 – 8, 1968 – 9, 1969 – 70; annual).

GHANA

Central Bureau of Statistics:

External Trade Statistics (1967, 1968, 1969, 1970, 1971; monthly).

Annual Report on External Trade (1966 – 8).

IVORY COAST

Direction de la Statistique, *Statistique du commerce extérieur de la Côte d'Ivoire* (1967, 1968, 1969, 1970, 1971).

Ministère de l'Economie et des Finances, *Statistique du commerce extérieur de la Côte d'Ivoire: imports et exports* (1971).

LIBERIA

Bureau of Statistics, *Manual for the Collection of Foreign Trade Statistics for Use on and after 1st January 1968.*

Department of Planning and Economic Affairs:

External Trade of Liberia: Exports (1968, 1969; annual).

External Trade of Liberia: Imports (1968, 1969, 1970; annual).

LIBYA

Census and Statistics Department:

Imports by Oil Companies (1969–70; annual).

Summary of External Trade Statistics: Listings of Imports – Exports, Re-exports (1970, 1971; annual).

External Trade Statistics (1968, 1969, 1970, 1971, 1972).

MADAGASCAR

Institut National de la Statistique et de la Recherche Economique (INSRE), *Statistique du commerce extérieur* (1967, 1968, 1969, 1970, 1971).

MAURITIUS

Customs and Excise Department, *Quarterly Statistical Report* (1967, 1969, 1970, 1971).

NIGERIA

Ministry of Commerce and Industry, *Nigeria Trade Journal* (1967, 1971, 1972).

Federal Office of Statistics:

Nigerian Trade Summary (1967, 1969, 1970, 1971).

Review of External Trade (1967).

SENEGAL

Direction de la Statistique:

Exportations, 1er semestre (1971).

Importations, 12 mois (1971).

SIERRA LEONE

Ministry of Trade, *Trade Journal* (1967, 1971, 1972).

Central Statistics Office, *Quarterly Trade Statistics* (1967, 1968, 1969).

TANZANIA

Ministry of Commerce and Industry, *Tanzania Trade and Industry* (1967, 1968, 1969, 1970; quarterly).

TOGO

Direction de la Statistique:

Annuaire des statistiques du commerce extérieur (1969, 1970; annual).

Commerce extérieur: exportations – importations (1971; quarterly).

TUNISIA

Institut de Statistiques:

Annuaire statistique du commerce extérieur (1969, 1970, 1971).

Statistiques du commerce extérieur (1969, 1970, 1971; 1967 and 1968 published by the Secrétariat d'Etat au Plan; annual).

UGANDA

Statistics Division, *Monthly Trade Bulletin* (1969, 1970, 1971, 1972).

UPPER VOLTA

Ministère des Finances, *Commerce extérieur: balance commerciale de la Haute-Volta* (1967, 1970).

ZAIRE

Institut National de la Statistique, *Bulletin trimestriel du commerce extérieur* (1969).

ZAMBIA

Central Statistical Office, *Annual Report: Statement of External Trade* (1967, 1968, 1969, 1970; annual).

CHAPTER 12

National Accounting Statistics

Broadly speaking, the core of most statistical programmes today is a system of national accounts. Such a system was first constructed to quantify the Keynesian model for unemployment in industrial countries and then adapted, in particular through the production of constant price series and sector output totals, to meet the increasing concern with economic growth during the 1950s and 1960s.

Essentially, national accounts present the records of economic transactions of an economy in a systematic manner and show the relationship between the various components of the economy. Economic transactions cover all the activities of an entity (households, governments, firms and financial institutions) that are of an economic nature (production, consumption, distribution, saving and accumulation, and foreign transactions). These economic transactions of all the entities in an economy can be presented in the form of accounts.

In Africa, the national accounts use the French (Courcier) and the British systems of national accounts.[1] The *Survey of Economic Conditions in Africa* remarks:

> Like much else in Africa, the major systems of national accounting presently in use in African countries reflect colonial experience; and systems in vogue were largely determined by historically accidental distribution of colonial power. Thus in the majority of the French possessions, the French (Courcier) system is in use; in countries where English is an important Lingua Franca a more or less modified version of the System of National Accounts (SNA) is prevalent. (ECA, 1970)

Accordingly, countries that have been under French or British influence in national accounting tend to adhere to the types of economic framework of accounts characteristic of these countries, and this has resulted not only in some differences in definition, but also in differences in the coverage and evaluation of items such as non-market production.

The United Nations (UN) System of National Accounts (SNA) attempts to reduce the differences between the two systems by incorporating commodity flows and specifying the resources and uses of commodities (United Nations, 1968). Despite the development of the new UN SNA, which all the countries of the UN, including African countries, are trying to adopt, national accounting in Africa is still at a very preliminary stage and rarely goes beyond the construction of the gross national product (GNP) and expenditure account and the balance of payments. Methodological problems still remain. The SNA does not make provision for accounts showing the activities of foreign companies (an omission that reflects its origin in countries where these are relatively unimportant). Such accounts can certainly be grafted onto the SNA, but its emphasis is wrong. Reflecting the practice of the industrial countries, it focuses attention heavily on the main tables, especially the gross domestic product (GDP), and the international agencies reinforce this bias by requesting national statistics offices to provide data for aggregates long before that preparation is defensible, resulting in figures that are little better than random numbers.

We shall therefore narrow down the subject of this discussion to national income or accounts data and, as far as possible, concentrate on what seems to be their inadequacies and biases both in their provision and their use, as well as some presentation of the conceptual framework underlying national accounts and the difficulties of measurements, rather than on national accounts as such.

SOURCES AND METHODS OF DATA COLLECTION

As pointed out earlier, the national accounting statistics in African countries relate basically to the tables relating to GNP, foreign trade, government revenue and disbursements, and some information on consumption. The UN Economic Commission for Africa (ECA), in its *National Accounts Statistics in Africa*, has classified the African countries into four categories according to their stage of statistical development.

In category I, we have countries that have held population censuses, agricultural censuses and countrywide household-budget surveys during the past ten years. In this category it is usually found that surveys on industry, employment and earnings and, in a few cases, surveys of distribution have also been conducted, in addition to a regular flow of data on external trade, balance of payments, government accounts and prices being available.

In category II, we have countries that have held population censuses and extensive rural and urban household-budget surveys, but have not attempted an agricultural census on a national scale. It is usual to find in this category of countries that industrial and employment surveys have also been held, in addition to reasonably good data on prices, external trade and government accounts being available.

In category III, we have countries that have held neither population nor agricultural censuses, but have conducted countrywide demographic sample surveys and localised agricultural household-budget surveys, annual data on external trade and government revenue and expenditure being generally available.

Finally, in category IV, we have countries for which statistical series are largely limited to those on external trade and government accounts and some statistics on prices, with perhaps the addition of one or two household-budget surveys of a sporadic nature. For this category of countries it is usually difficult to make valid estimates of national accounts.

No attempt has been made by the ECA to list the countries into different categories, as the transition of any country from one category to another is smooth and imperceptible.

CONCEPTS, DEFINITIONS AND THE STRUCTURE OF NATIONAL ACCOUNTS

Concepts and Definitions

First, a series of definitions. The factors of production of a country, during a given period of time, produce a certain level of output of goods and services. The *gross national product* (GNP) is therefore defined as the market value of the total output of goods and services produced by a nation's economy. *Gross domestic product* (GDP), on the other hand, is GNP less payments of wages, interest, rent, profits and other income items paid abroad. Nations in which foreigners have heavy investment holdings or whose citizens have extensive holdings abroad may find the GDP concept a more revealing gauge of economic activity than the GNP. The sum total of output of goods and services produced, valued at the prevailing market prices, is called the *gross domestic output* at current prices.

Gross domestic output at current prices less the cost of raw materials and containers equals the *net domestic output* at current prices. GDP at current prices less the provision for depreciation equals the *net domestic product at current prices.* Net domestic product at current prices plus subsidies less indirect taxes equals *net domestic product at factor cost*. Net domestic product at factor cost plus the net factor income received from abroad equals *net national product* at factor cost, which is also called *national income.*

National income is defined as the sum of incomes that accrue to the factors of production of a given country during a given period of time before deduction of direct taxes. National income can be measured at current prices or at constant prices. In the case of the former, as the name itself suggests, current factor costs are applied; while in the case of the latter, base year prices are chosen for the computation. The discussion on the choice of the base year and that of the index to be used are outside the scope of this discussion. The constant price series is generally used for comparison over time. Personal income less direct taxes equals *personal disposable income*.

Finally, the primary concern of economic analysis is with output or *real* income per head rather than the total output or income for the whole country. When national income or output is divided by the total population of a country, we obtain *per capita income* or output per person. We therefore go further and define *real gross national product* and *real gross domestic product*. The real GNP or real GDP is the GNP or the GDP figure adjusted for current inflation. The *gross domestic product deflator* is a measure of inflation computed as a ratio of current to real GDP.

Structure of National Accounts

Second, a discussion of the structure. The concepts of income and product are inter-related and can be derived from one another. These concepts correspond to the totals of entries in different accounts for the nation. A simplified structure of the national accounts is presented in Table 12.1.

The five accounts presented in the table show the structure of a particular country in a simplified manner. Each account has two sides: credit and debit, corresponding to income and expenditure respectively. All the incomings are recorded on the credit side, and all the outgoings are recorded on the debit side of the account. Moreover, each entry is recorded twice, on the income side of the one account and on the expenditure side of the other account. Note that the income of one account is necessarily the expenditure of another account. Thus, the expenditure on goods and services incurred by the households (debit) forms the income (credit) of the production account. All items can be similarly treated.

These accounts also focus on the methods of estimating GNP. The credit side of the production account records the items to be included in the GDP when approaching by the expenditure method, while the items recorded on the debit side give the components of the production approach. We therefore explain these methods in the next section.

The main questions are: whether special provision needs to be made in the accounts for foreign companies, public corporations or the informal sector; which key sectors need to be shown separately (oil,

Table 12.1 *A Simplified Structure of the National Accounts*

(a) *GDP and Expenditure Account*

Debit	*Credit*
GDP at factor cost (i.e. factor incomes)	Consumption expenditure of households
Provision for depreciation	Consumption expenditure of the government
Indirect taxes	Gross domestic capital formation
Less Subsidies	Exports
	Less Imports
GDP at market prices	Expenditure on GDP at market prices

(b) *Household Account*

Debit	*Credit*
Consumption expenditure of households	Wages and salaries and other incomes received from the production account
Income taxes	Transfer income from the government
Miscellaneous fees	Transfer income from the rest of the world
Savings	

(c) *Government Account*

Debit	*Credit*
Consumption expenditure of the government	Government income from property and entrepreneurship
Transfer income to households	Indirect taxes
Transfer income to the rest of the world	Income taxes
Subsidies	Miscellaneous fees
Government saving	
Total disbursements	Total incomings

(d) *Saving and Investment Account*

Debit	*Credit*
Gross domestic capital formation	Provision for depreciation
Net capital formation abroad	Saving of households
	Saving of the government
Gross capital formation	Gross domestic saving

(e) *Rest-of-the-World Account*

Debit	*Credit*
Exports of the domestic country (i.e. imports of the rest of the world)	Imports of the domestic country (i.e. exports of the rest of the world)
Transfers to the government	Net capital formation abroad (i.e. borrowing by the rest of the world)
Transfers to households	
Expenditure	Income

tourism, etc.); and whether there should be geographical divisions corresponding to political or socioeconomic entities. It would help the long term planning of national statistical work to indicate what sort of considerations need to be taken into account in constructing a structure of national accounts for any particular country.

The eventual accounting structure should, however, depend on a series of factors that determine the demand for statistics and their supply:

(1) *Geographical*: the area and diversity of a country; and the degree of urbanisation.
(2) *Political*: the long term objectives of the government and the interests of those outside the government in alternative development strategies; the degree and nature of government involvement in the economy; and the extent of devolution of powers to provincial governments.
(3) *Administrative*: the quality of administration, especially the number and quality of economists and statisticians; and the extent of data-gathering.
(4) *Economic*: the importance of foreign trade and the degree of concentration on a few export commodities; the proportion of economic activities covered by large units; and the relative size of agriculture.

These determinants of what sort of structure would eventually be appropriate appear so numerous that a very large number of structures would have to be considered. But the various characteristics are themselves associated; thus, small countries are generally more dependent on foreign trade or two export commodities.

METHODS OF ESTIMATING NATIONAL INCOME

There are three methods of estimating national income. These are: production or output method, income method and expenditure method.[2] These methods give the same results, and these can be explained by taking a simplified example. Suppose that a class of students goes for a picnic and they take with them bananas, oranges and apples for eating. We want to know the total number of these items. One way would be to approach the source of supply (i.e. the shop from which these items were purchased) and inquire about the total supply. The other way to find out the total quantities would be to approach each student and find out how many of these fruits were consumed by each student and how many are still left with him. All the three approaches would arrive at the same total. These approaches correspond to the output method, the income method and the expenditure method respectively of estimating national income. If we approach all the production units in the country and total the goods produced and the value of the services rendered by the service sector, we will get national income from the output method. If we approach each individual in the country and add up their total incomes, we will get national income from the income method. Further, if we estimate the total consumption expenditure of all the entities in an economy and add to it the total capital formation, we will get the national income from the expenditure method.

In actual practice, we cannot estimate the contribution of all the entities by all the three methods of estimation in Africa. The methods followed depend on the nature of the activity under consideration. For instance, the output method is best suited to the field of estimating output of agriculture, manufacturing and the like, where identification of outputs and inputs is possible. In other cases, where such an identification is difficult, as in the case of most of the services, the income method is applied. The expenditure method is applied when the data on the expenditure on various items either have been collected through household and other surveys or are available through the government records.

In agriculture, most of the African countries apply the production method. The information on production of various components of agriculture is collected through the marketing reports, sample surveys, crop-reporting, and so on. Several assumptions have to be made to collect the output figures of agriculture, in particular the unmarketed products, which constitute a substantial portion of the total agricultural output. In Nigeria, the estimates of farm products such as green vegetables, tomatoes, oranges, bananas, sugarcane, coconuts, pawpaw and pineapple are based on inquiries into the income and expenditure patterns of lower and middle income households, supplemented by export data. The output of these crops can be evaluated with the help of rural-market farm prices collected by the Federal Office of Statistics. The value added is obtained by deducting seed inputs from the gross output and is estimated at the constant prices of 1962–3. Since it is not possible to cover all crop items, a sample is usually taken to cover the remaining ones.

Estimating the contribution of livestock in the GNP of Nigeria, the output of hides and skins, meat and milk, may be estimated separately. The estimates of hides and skins are based on the export data. However, a deduction is usually made for the number of hides imported into the economy. The total output of hides is estimated by deducting 25 per cent of the quantity of net exports to allow for local consumption and is evaluated at f.o.b. prices. The only input taken into consideration is the transport cost. For 1969–70, the estimates published by the Federal Office of Statistics were based on the reported number of cattle slaughtered.

The data on cattle hides can be used to estimate the production of beef, goat meat and mutton by applying average carcass yields. The estimates of pork output are based on the official figures of slaughtered pigs.

For estimating the output of milk, a cattle population of 9·6 million in 1958 – 9 and an annual net increase of 200,000 heads are assumed. The output of milk is based on the assumption that the annual milk yield per cow is 60 gallons (274 litres). Milk output is evaluated at the retail price of sour milk at Kaduna. Similar assumptions can also be made for estimating the output of poultry and fishing. But these assumptions can be very unrealistic[3] at times. For example, as a result of pressure from international (or in some cases, bilateral) agencies on national statistical offices to provide data (which may not be available) to fill in returns, or to provide information to back an aid request, estimates have been produced (e.g. for food production or services) that have been neither needed by local policy-makers or analysts nor been of any defensible quality professionally.

The output method is used to obtain the output of mining and quarrying, which includes oil, metalliferous ores and coal. The annual reports of the Federal Ministry of Mines and Power have been used to estimate the production and f.o.b. prices of crude oil and gas and metalliferous ores. The Petroleum Division of the Ministry of Mines and Power gives information on crude oil and gas, while its Mining Division gives information on metalliferous ores. Information on the output of coal is derived from the annual reports of the Nigerian Coal Corporation.

The output method is used to obtain the estimates of the manufacturing and crafts sector. In the case of manufacturing, the information on the values of output and intermediate consumption can be obtained from the survey of manufacturing industry conducted by the Federal Office of Statistics. Crafts include the activities of those engaged in the production of durable goods (e.g. clothing and footwear) in their own houses. Okigbo's (1972) value-added data for 1957 were based on the average imputed earnings, and a year-to-year growth rate of the contribution to the real product of 1 per cent was assumed. These estimates were at 1962 – 3 constant prices.

In the case of building and construction, we can apply either the income method, by estimating the wages and salaries paid to the persons involved in construction, or the expenditure method, by estimating the material used in the construction of houses. In African countries, where in villages and in towns people still live in mud houses, the estimates are based on the estimated population using the mud houses, working out average life span of such houses and some estimates or assumption of the value added. For instance, in Nigeria, it is assumed that such houses have an average life span of thirty years, and 60 per cent of the value of building is regarded as the value added.

In the case of other buildings, which include concrete houses and buildings, the expenditure method is used. The cost of cement is used as an index for estimating the total value added.

In the case of electricity, all the three methods can be applied in estimating the contribution of this sector. Kilowatts produced give an estimate of the output side; kilowatts consumed give an estimate of the contribution of this sector from the income side. In most African countries, the records of the electric companies are available with regard to the production and payment of salaries. However, data on the consumption of electricity are scarce.

In the case of services, transport and communication and government administration, health and the like, the easiest way is to use the income method for estimating the contribution of this sector. In Nigeria, the income method has been used to estimate the value added by the services sector. As far as the data on salaries and wages paid to government employees are concerned, these are available through periodic records published by the government. However, the salaries paid by the private sector need indirect calculations, and hence several assumptions have to be made. The above analysis clearly shows how to estimate the output generated in the various sectors of the economy and the sources of information in African countries.

OTHER SUPPORTING DATA

In a number of African countries, some statistics are also available on other aspects of national accounts. These include consumption estimates, the balance of payments and saving statistics. These data are very useful in preparing national accounts and cross-checking the various data recorded.

Consumption Data

Consumption can be estimated by four methods namely: the commodity flow method, the physical quantities of goods purchased, the sales statistics, and the family budget data. The commodity flow method approaches consumption from the output side. The output of various goods is taken into consideration by making allowances for the uses of output as raw material, for capital formation, and for export and import. In order to estimate consumer purchases by this method, one needs to distinguish output by final use and to have information on transportation costs, distribution make-ups and foreign trade.

The other method, estimating the actual quantity of goods purchased, is very useful in African countries, where a large proportion of consumer goods is obtained by multiplying the area shown by the estimated yield. The output data thus arrived at is multiplied by the prices prevailing in the nearest local market. This method can predominantly be applied to the consumption of agricultural goods.

The third method, collection of retail sales statistics, is not useful as applied to African countries, for retail sales statistics are either not reliable or very scarce.

The fourth method of estimating consumption is the household budget data. Generally, sample surveys are required to construct the consumption estimates, and this method is widely used in African countries for building up consumption estimates. However, since the resources at the disposal of most African countries are limited, these surveys[4] are not conducted annually.

Gross Domestic Saving and Gross Domestic Investment

Savings are generally derived as a residual item of income and consumption. One can, however, estimate them indirectly by deducting the net inflow of capital from the gross domestic investment. Alternatively, one can look at the various components of savings: that is, the household savings in the form of deposits, provident funds, insurance, purchase of equities, and so on; government savings, given in the records of the government; and the enterprise savings, as shown in the income and expenditure accounts of a business. Since the national income data and the consumption statistics are based on various assumptions, then, of course, to derive saving as the residual of two items would in most African countries simply compound the errors of both. The method whereby sources of saving are located would be a better alternative. The deposit data of the households can be taken from the statements of the banks. Similarly, the other components of the household financial savings are published in the reports of the central banks of most African countries. Government savings can be taken from the government budget data; and the company's savings, which are the sum of the undistributed profits and the provision for depreciation, can be obtained from the income statements published by companies from time to time.

The weakest estimate would be the non-financial savings of the households. Households, in African countries, do not have the habit of depositing their savings in banks, except perhaps in the case of employees in wage economy and urban centres. Moreover, banking facilities are not always available in the villages and towns in the rural areas. Further, some proportion of the savings may be directly invested. For instance, for repairing his house, a person may directly use his income and may not borrow from a bank. Although the types of savings described can be used for building up reliable estimates of savings, the procedure presents a very difficult task.

There is, however, another problem. An aggregate account of savings and investment would be particularly liable to mislead the public, and even planners and policy-makers, because it could encourage them to assume that the total savings could be freely allocated to various forms of investment. In practice there is little flexibility. Those who save decide on how the savings can be used; thus, corporate savings are likely to be used for reinvestment in the same sector, private savings are often used for luxury buildings, and much of the inflow of capital may be destined for specific purposes, whether it originates in multinational corporations or aid agencies.

The third method is to deduct foreign investment from the gross domestic capital formation. The development plans of many African countries give data on capital formation, divided into fixed assets and the changes in stocks, as well as foreign capital. The reliability of savings statistics derived by this method will largely depend on the reliability of the investment data. Therefore, a detailed analysis of the investment statistics is essential.

In Nigeria, the Federal Office of Statistics publishes data on the gross fixed-capital formation by type of capital goods, both at current prices and at 1962 – 3 prices. The types of capital formation considered are: residential buildings, non-residential buildings, other construction except land, land improvements and orchard development, transport equipment, and machinery and equipment. As in other cases, the data on some components are more reliable than others. The information on the change in stocks can be obtained from the census of production data. The change in stocks of the entities not covered by the census of production would pose serious problems.

Export and Import Statistics

The data on exports and imports are published regularly by the central banks of African countries. Although the data are supposed to be reliable, this is, however, far from being the case, as far as imports are concerned. Smuggling goods into countries is widespread, thus undervaluing the official data on imports. In Nigeria, for instance, there is an official ban on the importation of many commodities, but these commodities can be found everywhere in the market.

Finally, commodity flow statistics tracing key primary products through their processing and distribution are of major importance. But the data on the distribution service is unsatisfactory, although the conceptual problems are less daunting. The level of disaggregation is poor in that it is difficult to follow through categories of goods by categories of outlet. If one cannot construct tables of commodity flows, or accounts of the key sectors, one cannot properly estimate the item in the main tables at all.

CONCEPTUAL AND MEASUREMENT PROBLEMS

The gross output in the standard national-income terminology is defined as the total turnover of goods multiplied by their respective unit prices. When we want to apply this concept to developing countries, we are confronted with the problem posed by unmarketed products, for which neither the estimate of quantity nor the unit price exists. For instance, the farmer does not supply all his produce to the market. Some of the output he consumes himself, and some of it he barters away (he purchases other goods in exchange for the output that he has produced). He may pay the barber by giving him maize for cutting his hair. The extent of the consumption of goods produced by the producer himself and that of barter are therefore substantial in the agricultural sector. There are no reliable estimates of the extent of non-monetary transactions in developing countries. A question therefore arises: if the problem is that serious, how can we solve it? We can approach the problem in a different way and apply indirect techniques for estimating the value of the unmarketed products.

The method involves three steps. (1) First, we must find out the acreage cultivated either from the village head or by conducting surveys, then find out the average yield (output per acre). By multiplying the total acreage with the yield, we can have a rough estimate of the total output. (2) The second step involves finding prices of the unmarketed produce. We can find out the prices of similar produce in the nearest local market where, presumably, the unmarketed items may have been sold. (3) Once we have obtained output and prices, by simply multiplying the two together we will obtain the value of the output. From this output we deduct the value of the marketed produce to estimate the value of the unmarketed produce. As an example of how the estimation of agricultural production has been undertaken in Ethiopia and illustrations of techniques, basic assumptions and the reliability of the data, Abraham (1972) has written:

> So far as can now be determined, the 'actual' level of cereal production was originally fixed for the year 1961 (there are indications, however, that the 1961 figure can be traced to a still earlier estimate for 1954 or 1955). Using the best fragmentary

evidence available, including the results of some small-scale family budget surveys (mainly for Addis Ababa), the annual consumption of all cereals taken together plus ensete (false banana) was put at 180 kg per person. An estimate of the total gross production of cereal crops was then arrived at by multiplying per capita consumption by the population and allowing for wastage and other losses, grain used as seed, etc. This total was in turn divided up according to type of crop using proportions suggested by the experience of the Ministry of Agriculture and the Planning Board. For the next few years these 1961 benchmark estimates were carried forward using a rate of increase somewhat above the estimated rate of population growth.

In December 1967 when the Third Five Year Plan was being readied, a committee of agricultural experts was informally convened to review the output estimates and to set plan targets. Making use of new but still incomplete information on areas under various crops collected in the first round of the National Sample Survey (NSS), the production series were substantially revised starting with 1963/64 to form a fresh basis for projecting output and a set of output growth rates was agreed upon which still serves as the basis for the estimates of crop production. Teff production was scaled down by a full one-third and sorghum by a quarter, wheat was raised 150%, barley 70% and maize 10%.

The second conceptual problem relates to the identification of the type of income or the industrial origin of income. The incomes in developing countries are of mixed character and do not lend themselves to classifications. It is difficult to find out the wage income, profit income, rental income and interest income in these countries. A person carries out different activities at the same time, and the income that he gets is a mixture of two or more activities. For instance, a farmer may also be carrying out trading and helping somebody on another farm. The farm income, wage income and profit income is a mixed package that he receives for performing these activities, and it is hard to classify his income by sources.

Caldwell (1977), using survey experience from Nigeria and other African countries, has agreed:

It is difficult to establish the probable level of a peasant's income, because he has no employer and no wage scale, although the market price for staples and his area of cultivation may be a guide to receipts for staples. However, his supplementary income may be substantial, difficult to detect and almost impossible to assess. In Nigerian research employing survey techniques which yielded plausible answers to most responses and data that were consistent with others collected independently, we have not been able to show satisfactory relationships of provable accuracy between stated income and other characteristics of respondents, such as occupation or education.

McDevitt, after a long period of intensive survey work in rural Nigeria, concluded that most of the money data were of little value and pointed out that the rural businessman does not make a distinction between his household and his business economy. Taylor, working in rural Kenya, employed an unusual research strategy to show that there the likelihood of income being revealed declines as its source becomes less respectable or less derived from personal exertion and more from investment: statements of income derived from coffee-growing were more complete than on income from a share in a lorry let alone a share in a bar or takings from prostitution.

The failure to disclose all earnings, investment and expenditure to interviewers is conventionally explained as fear of taxation authorities; yet our experience in Africa suggests that a much more potent fear is that the responses will be overheard by relatives and others who have never been certain of the amount of money dispensed by the respondent, and who have claims on his care may have received different assistances from that given to others. This is not evidence that such persons dislike expenditures of this type and will increasingly move to reduce them; rather is it an attempt to minimize competition, jealousy and envy.

Mueller attempts to circumvent the inadequacy of money data by employing an approach based on labour inputs into the growing of staples. This has two major weaknesses. First, it underestimates the complexity of peasant economies, which is particularly unfortunate in the context of her type of analysis, because a particular high proportion of children's inputs is into work other than the production of staples and because the question of value of children's work is likely to become a real issue in modernizing economies where diversification is taking place. Secondly, it is based upon measurements of labour inputs and consumption, which are suspect, at least in the case of the latter where none are clearly derived from adequate measurement in the developing world.[5]

Bondestam (1973), also commenting on the general reliability of income statistics, has written as follows:

Fairly reliable income data can be collected from large establishments, whereas income data from small enterprises, like shops, handicraft business, etc., are less reliable and also difficult to obtain. Income data for a majority of the population – farmers, livestock breeders and other own account workers, and those under- and quasi-employed – are extremely unreliable. The income estimates of these groups are also affected by seasonal variations in income, i.e. the estimated income depends on at what time of the year the question was asked (survey was conducted), and the data can therefore in most cases not be used to derive the annual income of the respondents. Moreover, the respondents' interpretation of a vague question like 'how much did you earn last month?' depends on various factors, such as the means and magnitude of income, the risk of being imposed heavier taxes, and his willingness to reveal his private economy in general.

This problem has become more and more realised in the planning of surveys dealing with family economy, and new approaches are made. One can either derive the income from knowledge about production (especially farmers) and occupation (from which the wage can sometimes be estimated), but this method is difficult to apply due to the problems of estimating production and in finding out all the various sources of income of a family. The other possibility is to estimate the consumption (house rent, consumption of food and clothes, etc.) and savings (or debts), but even this method lacks a tolerable degree of reliability, due to the difficulties in measuring the price of consumption goods, the interviewee's difficulties in remembering all his/her expenses, the difficulties in measuring the value of gifts and the transference of money between families and the unreliable information one can expect to get on debts and particularly savings.

Unreliable income statistics are, however, not limited to the poorer part of the population. Even salaries of well paid businessmen in large enterprises, whose managements for various reasons may be interested to hide the true economy of these enterprises, are not always fully reliable. So, for example, a comparison between two large plantations in Ethiopia shows that the official average salary for a foreigner in one of them is about US $10,000 yearly, but in the other only US $4,500, a difference which is definitely unrealistic.

African income statistics are scarce, the data – if they exist – are often too sensitive to be accepted for official publication, and, in consequence, their reliability has not been much elaborated on by statisticians and economists in general.

When distributions are very heterogeneous or very highly skewed, averages became very misleading, particularly when comparing different groups. Bondestam (1972), giving an example of this from Kenya, has written as follows:

The monetary economy is concentrated to the main towns (and

large commercial farms) and especially to Nairobi. In 1970, 66·1% of all wages and salaries paid in all towns went to Nairobi – an increase from 63·8% in 1964. The concentration of capital in towns is reflected in the fact that average earning per employed town-dweller is 14 times as high as that of an average farm labourer. An average wage earner in Nairobi gets even more, namely 17 times as much as the rural labourer. However, there are large differences also comparing various rural regions: an average regular employee in small farms or settlement schemes earns twice as much in the Central Province and almost three times as much as the one in Nyanza Province. Of all wage earners, the average European earns more than 12 times as much as the average African, a figure which has only slightly decreased since 1964, when the ratio was 13/1. Earnings per capita have increased by 8·0% annually for Africans from 1964 to 1970, mainly due to the increasing portions of Africans in the higher income brackets (the proportion of Africans earning more than 40 pounds per month increased during the six years from 10·2% to 16·3%), whereas the average Asian had his income increased by as much as 9·0% yearly during the same time, although he earned five times more than the African already in 1964. The large variations in income by race are further reflected in the proportions in the higher income bracket: 1·4% of the Africans, 20·9% of the Asians and 54·0% of the Europeans earned more than 100 pounds monthly in 1970. Although the situation for women has improved remarkably since 1964, as many as 47% of the females, compared to 'only' 23% of the males, earned less than 5 pounds per month in 1970. Finally, adding regions, sexes, and races together, one finds that half of the wage earners get less than 10 pounds per month, whereas more than 10% have a monthly salary of more than 50 pounds.

These examples have been given to show the tremendous regional, rural – urban, race and sex variations in wages and salaries. One may then argue that, even if there are variations within a country, the average income *per capita* can be used in comparisons between countries, as a measure of the magnitude of the level of living. Table 12.2, where the estimated income distributions in the cities of Nairobi and Addis Ababa are compared, may throw sufficient doubts even here.

If the last column referred to the total population instead of wage-earners only, the percentage figure for Addis Ababa on line '$1 – 99' would be even larger! From the comparison in the table, which does not need any comment, one can conclude that any comparison between countries with respect to average income *per capita* is entirely meaningless, unless accompanied by an income distribution of the respective country, and this is usually very hard to find in the available

Table 12.2 *Estimated income distributions in Nairobi and Addis Ababa.*

Monthly Income[a]		*Income Distribution*	
Kenya *(shillings/month)*	Ethiopia *($/month)*	Nairobi *(% of entire city population)*	Addis Ababa *(% of wage- and salary-earners)*
1-331	1-99	22.7	46.6
332-665	100-199	27.7	20.7
666-1,665	200-499	22.6	22.6
1,666 plus	500 plus	27.0	10.1

[a] One Kenya shilling equals approximately 0·3 Ethiopian dollars.

statistics. (The same type of discrepancies in the income distribution can, of course, be shown for many other African countries as well.)

The third conceptual problem relates to the structure of developing countries. Most of the economic activity is done on a small scale, and generally the accounts of output and input are rarely kept. Illiteracy also contributes to the non-recording of transactions. Due to these reasons it is difficult to apply the concept of value added, which is defined as the output less the cost of raw materials and depreciation. Bondestam (1973) has agreed and given an excellent example as follows:

> A major part of the population in Africa is outside the monetary economy, i.e. they live mainly on subsistence production although small surpluses of the produce may be sold on the market.[6] One study claims that 'less than 15% of the total farm product is marketed and the small surpluses yielded by subsistence cultivation are usually traded in markets within the traditional economy.' (U.S. Department of Commerce, 1967). The estimate of the agricultural output from these farmers is arbitrary and extremely unreliable. The value of the products used for own consumption can hardly be measured in monetary units, particularly in societies which are not yet integrated in the monetary economy. Moreover, the exchange value of the surplus production sold on the market is not measured by the labour time needed to produce it but the value is rather a function of supply and demand and of profits by middlemen.

In African countries, begging is not uncommon. One often comes across beggars at the filling stations, shopping centres, markets, and so on. Now the problem is: should we include their income while computing the national income or not? There are two sets of arguments. The one that asserts that their income should also be included in the national income is based on the social necessity to give to beggars. In advanced countries, the governments give social insurance benefits to destitutes; and since in African countries there is no such provision for them, their income should also be included in the national income. Those who oppose their inclusion argue that it is merely a transfer of one person's income to another person; and since we have already included the donor's income in the national income, the inclusion of beggars' income would lead to double counting and hence inflate the national income. The standard practice followed in such cases is to exclude these incomes.

Another problem relates to the contribution of the housewives. In developed countries housewives perform several functions at home, while their counterparts in developing countries go out to work and earn an income, and in turn pay for some of the functions in the home to be carried out by others, which they might perform themselves if they were not employed. Should we impute the contribution of housewives in developing countries and include this in the national income? If yes, then how should we estimate their contribution? Prest and Stewart (1953) have favoured their inclusion and suggested that the bride price should be taken as an indicator of their contribution. If we take bride price, which differs in the case of different brides, we would again be doing injustice, as the bride price in most cases is nominal and definitely less than the value of the services that a housewife will render for the whole of her life. Moreover, if we start imputing these services, there is no end to them. For instance, a man shaves at home instead of hiring the services of a barber. Should we include his services to himself also in the national income? The practice followed in national income computations is to ignore such imputations.

Finally, the existence and importance of various types of semilegal or illegal transactions make the task of quantifying economic activities difficult, if not impossible, in Africa. Some of these are:

(1)smuggling operations across frontiers and between regions within the country;
(2)transfer pricing by multinational corporations;
(3)transactions financed by 'black money' accumulated by evading taxes; and
(4)prostitution, illegal betting, and so on.

Given the size and importance of these activities in many countries, national accounts that do not take account of these operations are unlikely to be of much use. There are, however, practical difficulties in applying the conventional rules of statistics in measuring these operations.

STATISTICAL LIMITATIONS OF NATIONAL INCOME STATISTICS

The data relating to national accounts of African countries suffer from severe limitations. These limitations stem from two basic reasons. First, the concepts of national income and national product have their origin in the advanced countries with a specific socioeconomic background. These concepts may not suit the socioeconomic structure of the developing African countries, and they can be labelled as conceptual problems. The second set of problems relates to the collection, tabulation and analysis of data and also the reliability of data. These problems can be called statistical problems. So far we have discussed the conceptual problems. We shall now discuss the statistical problems.

The data on quite a number of items do not exist, and whatever data we get are based on several assumptions and are thus highly unreliable. For example, in the case of estimating GDP of Nigeria, several assumptions have to be made with regard to output and prices of the product. In the case of output of eggs, it is assumed that 70 per cent of the poultry are hens, that a hen lays forty-eight eggs per year and that about 33 per cent of the eggs are hatched. A price per dozen at 2 – 2·5 shillings is applied (Nigeria, 1968, p. 10). Similarly, in the case of fishing, no inputs are taken into account. Value added is taken to be equal to output (p. 11). Owing to the lack of data, the Lagos prices for fresh fish are applied throughout the federation (p. 11). In fact, these assumptions are made in most of the calculations.

Killick (1978), writing on the national accounts and other statistical sources used in his analysis of Ghana's economy, has commented:

> Official estimates of Ghana's national accounts begin with 1955, although unofficial estimates have been made back to 1950. Rather unusually, they were derived from expenditure estimates, the first official estimates of the industrial origin of the GNP begin only in 1965 and no estimates have yet been attempted of the national accounts by factor incomes. At the time of writing, published estimates do not extend beyond 1969, which is a serious handicap in discussing the post-Nkrumah record.
>
> Various components of the estimates of expenditure on GNP are subject to substantial margins of error. A basic weakness of them all is that there is no meaningful series on the production or consumption of local foodstuffs – an item which alone is believed

to comprise 30–40 per cent of the total GNP. In an attempt to overcome this deficiency, a nation-wide sample household expenditure survey was conducted in 1962 and from this it was possible to obtain an estimate of national food consumption. The expenditure estimate for this item in the national accounts, as published in the annual Economic Surveys, is said to be based on the assumption of a constant real per capita consumption of local foods (although the published figures do not always seem to tally with this procedure) and, for the current-price series, is inflated by a national index of local food prices. Since there are several indications that food production and consumption failed to keep pace with the growth of population in the first half of the sixties, and perhaps in some later years too, this procedure, besides being arbitrary and useless for showing year-by-year variations, probably gave an upward bias to estimates of private consumption and GDP in this period.

Other items in the series are also subject to large error margins, although they are less important in relation to total GDP. Estimates of the construction component of fixed capital formation are largely based on imports of construction materials and, in the case of rural communities, assume the housing stock to expand at the same pace as population. The item for changes in stocks has a very limited coverage, being restricted to cocoa and the inventory position of a few major trading companies. There is also known to have been considerable over-estimation of public consumption, due to double-counting of the consumption expenditure of various parastatal organizations, beginning around 1966, biasing the accounts upwards to a significant degree. Finally, the price indices used for constructing constant-price series were of dubious quality, especially for public consumption, capital formation and external trade.

Appreciating the weaknesses of their estimates, the Central Bureau of Statistics (CBS) undertook a major revision of the national accounts, concentrating on the development of a series of accounts by industrial origin but also providing expenditure estimates consistent with those by industrial origin. To date, this new series only covers 1965 to 1968. Prepared in meticulous detail, this revised set of estimates must be regarded as a major improvement on most figures in the old series, although its authors were still gravely handicapped by an absence of tolerably accurate data on foodstuffs production. The new estimates showed total GNP to be about 15 per cent smaller than had previously been thought, taking the four years as a whole. As well as having a smaller absolute size, these estimates (which I call the Singal/Nartey estimates, after their authors) showed that the constant-price GNP fell a little over the four-year period, as compared with a 6 per cent rise indicated by the earlier series.

The substantial differences between the two sets of estimates and the brief span of the Singal/Nartey accounts created considerable problems for this writer because he needed a series that would cover at least the whole of the sixties. It was therefore necessary to reconcile the Singal/Nartey figures with the earlier ones, and the results are set out in Tables 4.A and 4.B, with the methodology explained in the table notes.

A feature which greatly assisted reconciliation of the two series was that for 1965, which was used as the base, the proportional composition of expenditures on the GDP was very similar in the two cases. Total consumption was 90·8 per cent and 91·7 per cent, fixed capital formation 16·9 and 18·1 per cent and domestic savings 9·2 and 8·3 per cent. Moreover, they both showed the share of private consumption and of fixed capital formation to decline over the period, with the share of public consumption growing (but more slowly in the Singal/Nartey figures) and domestic savings displaying similar trends.

Given these points of agreement and the fact that the general trends indicated by the series are consistent with independent indicators of the same features, the figures presented here are probably reliable enough to portray broad medium-term trends in the economy, which is the chief use made of them in this study. However, it is important to have an appreciation of the serious deficiencies of the data, so as to avoid placing weight on any set of numbers which will not bear it. For this reason, for example, particularly cautious use is made of the data on capital consumption and on Net National Product (NNP) for the depreciation estimates must be judged among the least reliable, even though the resulting estimates of net saving and investment are generally consistent with what one would have expected.

Despite all the weaknesses mentioned and the difficulties of reconciling the two sets of estimates, there is, in my judgement, little doubt that the broad trends discussed in the main body of this chapter did really occur, even though no great precision can be claimed for their measurement.

These deficiencies are symptomatic of the general problems of reliability of statistics in Africa. Enabor (1976) has written:

In Nigeria, data on national income and population are incomplete or inaccurate and generally unreliable . . . Nigeria's Gross National Product grew, on the average, at 5·5 per cent per annum in the 1960's (Lewis, 1967). Available evidence indicates that the GDP growth rate has tended to fluctuate from year to year, though on a generally rising trend (Aboyade, 1969). This may partially reflect statistical inaccuracies or changes in coverage or definitions. The Second National Development Plan (1970–74)[7] adopted 6·6 per cent per annum target growth rate for the GDP. Recent evaluations (Central Planning Office, 1973)[8] show that the average growth rate of GDP since 1970 is about 10·0 per cent. It is the objective of the Third National Development Plan to raise the Gross Domestic Product of Nigeria by about 9 per cent per annum in real terms.

It should be stressed that the recorded annual growth rates of GDP in Nigeria since 1970 are unusually high and have been made possible by the expansions of the petroleum industry. The growth rates cannot be maintained into the distant future. They need to be modified for purposes of long term planning. In addition, the growth rate of GDP in Nigeria clearly overstates[9] the growth of Gross National Income, GNP and purchasing power of Nigerians in view of foreign capital and management in the petroleum industry.

USES OF THE NATIONAL ACCOUNTS STATISTICS

National accounts present data in a coherent form covering all aspects of an economy, and as such they are very useful for economic planning and formulation of economic policies. The series of data on the GNP and net national product gives an idea of the overall productive capacity of the economic system; and when it is studied in relation to labour and capital, it shows the productivity of the factors of production. If it is studied in relation to consumption, it represents the standard of living of the people. However, if it is studied in relation to the distribution, it can show the equity of the economic system.

The *per capita* income is useful for indicating in time series the relationship between population and income or product growth. This is obtained by expressing the statistics of the main aggregates on a *per capita* basis to provide a standard. The exercise has statistical usefulness only if the following limitations are borne in mind. First, the national average should be accompanied by some data on the distribution of income, unless we can assume that the distribution of income in the two countries is approximately the same. There is a great deal of difference in terms of real living standards in a country with a *per capita* income of, say, US $400 per annum where income distribution is approximately equal, and one where 20 per cent of the population have an income per head of US $1,500 and 80 per cent an income of US $125. Second, in considering living standards, changes in the distribution of income over time may have as great an effect as changes in the gross product.

The formulation of the monetary and fiscal policies requires a strong data base. Operationally, the monetary authorities are interested in the availability of finance for each sector of the economy, and need to see a systematic presentation of the sources and use of funds, and the intermediation of financial institutions. Indeed, most of the activities of financial institutions that concern the monetary authorities would not appear anywhere in the national accounts if it were not for the flow-of-funds analysis. So this has rapidly been adopted as the appropriate framework for assembling statistics of financial assets and liabilities, and provides the criteria by which decisions can be taken about new statistical requirements.

Flow of funds has since won almost unanimous approval as a necessary part of the general body of national and international statistics. Started as an interlocking set of financial accounts for the main sectors, and providing an analysis of intersectoral flows in financial assets, it has more recently been extended to encompass sector balance sheets also. It is, however, expensive in its use of statistics, since it takes the form of a complete matrix in which a great many of the cells have to be filled, by guesswork if in no other way, before the analysis can become useful.

The information on the taxable capacity of individuals, their expenditure on various items of consumption (luxuries and necessities), the proportion of monetised consumption and the items recorded in national accounts are useful in formulating monetary and fiscal policies.

One can analyse also the impact of the government transactions on the economy as a whole by studying the government accounts together with the other component accounts. For instance, if the government of any country is adopting the deficit-financing policy, one can estimate whether output will rise enough to cope with the excess demand created by the deficit-spending of the government. If the productive capacity fails to correspond with the excess demand, it will lead to inflation.

Most African countries want to develop faster than their resources permit. Saving (household, government and enterprise) generally runs short of investment, due to overambitious planning. Consequently, these countries look for foreign aid to fill up the gap between domestic saving and domestic investment. National accounts portray the significance of foreign borrowing explicitly. Moreover, it is through the national accounts that we know the relative importance of each sector in financing economic development.

National product, being the measure of economic progress, can be used as an indicator for comparisons over time and over space. Each of the comparisons has limitations, as changes in the output basket over time and over countries, value of money and a host of other problems are faced while carrying out these comparisons. All these problems are taken care of by the economists when such comparisons are made.

The other use of national accounting stems from the fact that it is an accounting system. Any error can be easily detected because such accounts are prepared on the double entry system. But this advantage can only be reaped when the statistical system is advanced and there are precise calculation procedures. In African countries, the item 'statistical discrepancies' is included in the accounts. The cross-check provided by such an item in the accounts is therefore of a limited value.

CONCLUDING REMARKS

National accounts are prepared for a few years in most African countries. Nevertheless, the statistics on the GDP by industrial origin are compiled and published regularly. A beginning has been made, and sustained efforts are required in this direction. At present the data suffer from various limitations and are computed, barring a few items, by making several assumptions. It is still a long way before reliable data of national product become available. As far as the data of the government consumption of goods and services are concerned, they are fairly reliable. The same may be said of the export data. The official import data are underestimated to some extent by smuggling. Data on private consumption are still in their infancy. Savings and capital formation data have the same fate. All these data on national product, savings and investment, and private consumption are important for any meaningful planning at the national level. Since most African countries have launched plans for economic development, serious efforts are required to improve the quality and the quantity of the national accounts statistics.

It is therefore to the discussion of a few indications of gaps in the statistics for various sectors that we now turn our attention. On foreign enterprise, data on capital stock and detailed data on gross investment branch by branch are now lacking. In their absence, no clear indication of factor substitution between labour, skills and capital is possible. A detailed breakdown of data on intermediate inputs is also urgently needed. Comparative data must be published on foreign-owned and locally-owned manufacturing enterprises. Although much information is available, it is not easy of access.

On domestic enterprises the data available at present refer mainly to large firms employing over fifty workers. If progress is to be measured periodically, it is extremely urgent to collect regular statistics covering such matters as value added, average earnings, capital stock, capacity utilisation and employment, for firms occupying 5 – 10, 11 – 20 and 21 – 50 employees or other persons. The operation of a national tariff and protection policy also requires more information on the relative efficiency of, and the costs incurred by, different types of firms, or firms in different branches of industry.

A census of urban and rural small-scale industry is needed. Some of the data forms must be modified in order to obtain comparable information on employment by status and sex, branches of activity and capital investment. Such a census should both make it possible to judge the efficiency of the labour-intensive small-scale enterprises and indicate promising lines of action.

Even the data on distribution and services are unsatisfactory, although the conceptual problems are less daunting. The special capacity of the services sector to absorb labour is partly due to its greater opportunities for part time employment. Female and older workers who may not be normally available for regular and full time work go in for such activities as commerce and personal services requiring less physical strength and offering greater part-time opportunities. Thus, regular annual surveys of services and distribution should not only take account of part time workers in the total number of persons engaged but also give adequate information on the amount of time worked by part-timers. This type of information is not essential if income data are collected by sources of income and if they identify all types of activity (distinguishing part time from full time) that generate income. However, information on the amount of time worked makes it possible to take account of part-timers in full time equivalents. It also facilitates the inclusion or exclusion of unpaid family workers in figures of employment. For most purposes, part-timers and family workers cannot be given the same weight as full time workers.

On the problems of conceptual difficulties, mention has been made of the fact that published economic statistics refer almost exclusively to the output of the organised 'enumerated' or formal sector, implying that it has superior social value. Similar points can be raised about subsistence agriculture. One reason for the neglect of this sector may be the paucity of data about it, apart from a guess included in the estimates of domestic product. Yet, it is the starting point of a socially oriented development plan. The whole treatment of non-monetary output in the social accounts reflects an assumption that this type of activity is a sort of excrescence of the body economic, something for which an allowance can be added on (very approximately) after economic activity proper of various kinds has been estimated.

This is a reasonable assumption for statisticians in a developed industrial economy, where non-marketed output means mainly a little fruit and vegetable production, but it is a very different matter where a large part of all types of activity, not merely farming but also secondary production (milling maize, making clothes, building houses) and tertiary services (hair-cutting, laundering, transport, marketing), is carried out either within the family, on the basis of simple barter or in other ways not enumerated. Similar problems, which we have mentioned early in this book, arise over defining employment, unemployment and the economically active population. Statistical conventions are bedevilled by the coexistence of organised economic activity, activity that is unorganised but remunerated, and household activity carried on without payment, often all producing the same product. What is involved therefore is not just a switch of emphasis in the collection of data but a change in the conceptual framework.

Finally, we must mention that the statistical materials on which national accounts are based in Africa are of narrower coverage and of lower reliability than those of the developed countries. This raises problems of international comparison of the data. When, therefore, faced with statistics of a number of countries all on the same format, it is as well to remember that, although they appear to obtain comparable results, the reliability and coverage of the sources that go into the making of the aggregates may be different. It may need a great deal of practical knowledge and familiarity with the statistics of the countries concerned to assess the degree to which comparability is impaired by this problem.

NOTES: CHAPTER 12

1 See the section on the 'Structure of national accounts' for a full discussion.

2 In economies where production takes place for use rather than sale, income, product and expenditure are one and the same economic activity, not different aspects of the same total.

3 A director of statistics was asked at a conference how he justified the assumption built into his national accounts that food output per head of the population remained constant. 'Well,' the reply came, 'when I motor through the rural areas I don't see any sign that it changes.' See, for example, the discussion on national economic accounting in Dasgupta and Seers (1975, p. 28).

4 For a thorough discussion on the coverage and other details of data collection, see Economic Commission for Africa (1970), pt 2, table 2.

5 The references contained in the quotation are McDevitt (1975), Taylor (undated) and Mueller (1975).

6 According to one study, 66 per cent of the population belongs to the subsistence group, 54 per cent of peasants are in agriculture and 12 per cent are pastoralists and other non-agriculturists (e.g. collectors and hunters). For a thorough discussion, see, for example, Bondestam (1972).

7 See Nigeria (1970).
8 See Nigeria (1973).
9 This point was made when we attempted to define GNP and GDP. Because of foreign domination of the oil sector, national income is expected to be considerably less than GDP.

REFERENCES AND FURTHER READING

Aboyade, O. (1969), 'The economy of Nigeria', in P. Robson and D. A. Lury (eds), *Economies of Africa* (London: Allen & Unwin).

Abraham, W. I. (1972), *Ethiopia's Estimates of Cereal Production: How Good Are They?*, Staff Report No. 3 (Addis Ababa: Central Statistical Office).

Arya, P. L. (1976), *Social Accounting for Developing Countries* (New Delhi: Macmillan Co. of India).

Bondestam, L. (1972), *Population Growth Control in Kenya* (Uppsala: Scandinavian Institute of African Studies).

Bondestam, L. (1973), *Some Notes on African Statistics: Collection, Reliability and Interpretation*, Research Report No. 18 (Uppsala: Scandinavian Institute of African Studies).

Caldwell, J. C. (1977), 'The economic rationality of high fertility: an investigation illustrated with Nigerian survey data', *Population Studies*, vol. 3, no. 1, pp. 5 – 27.

Dasgupta, B. and Seers, D. (1975), *Statistical Policy in Less Developed Countries*, IDS No. 114 (Brighton: Institute of Development Studies, University of Sussex).

Economic Commission for Africa (ECA) (1970), *Survey of Economic Conditions in Africa* (Addis Ababa), pt 2.

ECA (undated), *National Accounts Statistics in Africa* (Addis Ababa).

International Bank for Reconstruction and Development (1976), 'Population *per capita* product and growth rates', in *Atlas* (Washington, DC).

Killick, T. (1978), *Development in Action: A Study of Economic Policies in Ghana* (London: Heinemann).

Lewis, W. A. (1967), *Reflections on Nigeria's Economic Growth* (Paris: Organisation for Economic Co-operation and Development).

McDevitt (1975), 'On the pragmatic side of migration fieldwork in Nigeria', paper presented at a seminar on Migration in Nigeria, Ile-Ife.

Mueller, E. (1975), 'The economic value of children in peasant agriculture', paper presented at the Resources for the Future conference on Population.

Nigeria (1968), *The Gross Domestic Product of Nigeria, 1958 – 59 to 1966 – 67* (Lagos: Federal Republic of Nigeria).

Nigeria (1970), *Second National Development Plan, 1970 – 74*, Vols 1 and 2 (Lagos: Central Planning Office, Ministry of Economic Development and Reconstruction).

Nigeria (1973), *Economic and Statistical Review* (Lagos: Central Planning Office, Federal Ministry of Information).

Nigeria (1977), *Economic Indicators* (Lagos: Federal Office of Statistics).

Okigbo, P. N. (1962), *Nigerian National Accounts, 1950 – 1957* (Lagos: Federal Government Press).

Prest, A. R. and Stewart, J. G. (1953), *National Income of Nigeria, 1950 – 51* (London: HMSO).
Taylor, F. (undated), Personal communication to J. C. Caldwell, Carlton University, Ottawa.
United Nations (1968), *A System of National Accounts*, Studies in Methods, series F, no. 2, rev. 3 (New York: UN).

MAJOR SOURCES OF NATIONAL ACCOUNTS STATISTICS

GABON
Service National de la Statistique, *Comptes économiques, évolution 1960 – 70, année 1968.*

GHANA
Capital Investment Board:
Annual Report and Accounts (1967 – 8; annual).
Investment Journal (1970, 1971, 1972; quarterly).

IVORY COAST
Ministère du Plan, *Les Comptes de la nation*, vol. 1 (1960 – 5), vol. 2 (1966 – 7), vol. 3 (1968).

KENYA
Economics and Statistics Division, *Input – Output Tables for Kenya* (1967).

LESOTHO
Bureau of Statistics, *National Accounts* (1966 – 7, 1967 – 8).

LIBYA
Directorate General of Economic and Social Planning; National Accounts Section (Tripoli):
National Accounts of the Libyan Arab Republic, 1962 – 71.
Estimates of Private Final Consumption Expenditure of the Libyan Arab Republic, 1962 – 9, supplement to the national accounts volume (December 1971).

MALAWI
National Statistical Office, *National Accounts Report* (1964 – 7, 1964 – 8, 1964 – 9, 1964 – 70).

MALI
Service de la Statistique Générale, *Comptes économiques du Mali* (1964 – 7, 1967 – 8, 1969).

NIGERIA
Federal Office of Statistics:
Estimates of Capital Formation (1969 – 70; annual).
GDP of Nigeria (1958 – 9 to 1966 – 7, 1958 – 9 to 1968 – 9).

SIERRA LEONE
Central Statistics Office, *National Accounts of Sierra Leone* (1963 – 4 to 1968 – 9, 1963 – 4 to 1969 – 70).

SUDAN
Department of Statistics, *National Accounts and Supporting Tables* (1966 – 7, 1969, 1969 – 70, 1970 – 1).

SWAZILAND
Department of Statistics, *National Accounts* (1967 – 8, 1968, 1968 – 9).

TANZANIA
Ministry of Economic Affairs and Development Planning; Bureau of Statistics (Dar es Salaam), *National Accounts of Tanzania* (1966 – 8, 1964 – 70).

CHAPTER 13

Socioeconomic Indicators

The purpose of this chapter is to summarise in a systematic manner the discussions in the previous chapters of this book. Our approach is not to repeat what has been said before (even though we may emphasise certain aspects of it) but to give a new orientation to the summary. The idea is to introduce the concept of 'indicators', a term that has become fashionable in recent years in describing and measuring development achievements and targets.

The degree and the rate of development are generally measured as the ratio of anything to the total population: for example, the gross national product (GNP) *per capita*, the number of doctors per population. But the unwillingness to define development in terms other than rates and ratios leads to severe misunderstandings on the true issue of underdevelopment and is therefore also one of the basic obstacles to a satisfactory economic growth and a reasonable distribution of income. Nevertheless, in this chapter, an attempt is made to define 'indicators' and to discuss their uses, problems of selection, construction of indicators, 'key development variables or indicators' such as the GNP, and the availability of socioeconomic indicators in Africa.

WHAT ARE SOCIOECONOMIC INDICATORS?

Indicators are surrogates for direct measurements, which approximate, represent or indicate the extent of the quality or property in question. The socioeconomic indicators thus relate to both the social aspect and the economic aspect of measuring development. Socioeconomic indicators thus include production, supply and disposition of goods and services, financial transactions, employment and unemployment, household income, savings and consumption, health, housing, social security, public order and safety. The socioeconomic indicators are therefore part of a multidimensional interdependent system, reflecting the pattern of development.

USES OF SOCIOECONOMIC INDICATORS

The uses of socioeconomic indicators are as many as one can conceive of, as they cover a very wide range of spheres. These uses can be for short run as well as for long run policy and planning and also for academic interest. The uses of some of the indicators are discussed in the chapters dealing with the topic, and for reasons of economy we avoid repeating them here. Broadly, however, we can say that the socioeconomic indicators are needed:

(1)to provide information about social and economic conditions and to measure changes in these conditions;
(2)to analyse and diagnose the conditions; and
(3)to formulate policies and assess their efficacy.

Economic index numbers and social indicators therefore play a large part in determining government policy. The influence that they have on many aspects of our life and society is probably greater than that of any other aspect of statistics.

SELECTION OF SOCIOECONOMIC INDICATORS

The selection of indicators is closely related to the uses to which they will be put. Nevertheless, the distinction between the use of socioeconomic indicators for analysing the situation and their use for planning is desirable. In the former case, the indicators measure a small part of what is more a complex historical process; whereas in the latter case, the focus is on specific areas of public policy or areas of social concern. A number of criteria are therefore used for the selection of indicators. These are theoretical as well as statistical. The theoretical or conceptual consideration that goes towards choosing an indicator is, first: how appropriate (i.e. how relevant and indicative) are these indicators to the measurement of social concerns for which they are intended? How great is their degree and power of summarisation, and how well are the series of indicators co-ordinated, linked and structured? Considerations on the statistical side refer to the reliability, accuracy, consistency over time, timeliness and feasibility of their compilation.

Appropriateness

To be appropriate to the social concerns or social policy, indicators should furnish measures of absolute levels or trends that conform with the aspect of the living conditions or the social service in question.

These measurements may be direct or indirect (proxy). The subject and the unit of measurement of the former should coincide with the subject matter of the study itself directly. For example, if the purpose of study or inquiry is population growth, then the ideal indicators are those which relate to the rates of new increase in population. This method is also referred to by social scientists as output approach.

The same result as above can be achieved indirectly by estimating the inputs or the other relevant series with which it has a high degree of correlation. For instance, in the case of measurement of health, one can take the expenditure on the health programme as proxy for improvement in health; or one can measure the level of education by the attendance in schools. Some observers feel that these substitutes for ultimate measures of the totality of a subject of social concern are likely to be better indicators of its trend than the absolute level. There are other observers who contradict it. But in African countries, the choice of direct or indirect measures of indicators is restricted by the availability of data.

Summarisation

The social indicators chosen should be able to summarise the information required without jeopardising the information on major facets of the concern needed for its assessment. This will keep the number of indicators of a study to the minimum. A social indicator may be designed to furnish the information on the whole or a substantial portion of a social concern. This can be done by aggregating data on the components – the direct approach. Alternatively, this can be done by the indirect approach through covering the key underlying circumstances, inputs or explanatory factors. However, in a number of cases the relevant information may not be available. For instance, if one wants to devise a comprehensive summary measure of the state of working conditions or the severity of criminal offences and victimisation, one is faced with the problem of judging correctly the strategic factors. One can overcome these limitations of the direct measure to some extent by adopting the indirect approach.

Accuracy

The degree of accuracy of the indicators depends basically on the statistics on which they are based. It is undisputable that in order to be useful the indicators must be accurate. Any gap or anomaly in the social indicator will reflect the weakness of the social statistics. The social indicators, direct or proxy, should be used after careful assessment of the statistics from which they are drawn or on which they are based.

Timeliness and Frequency

The indicators should furnish a timely review of the social and economic conditions prevailing in the country, so that the situation can be arrested. The indicators that are of major current use should be collected more frequently. The indicators that are of occasional interest can be constructed by benchmark data when the need arises. The indicators on the welfare of population, such as employment and unemployment, should be constructed more frequently, perhaps on an annual or bi-annual basis. At present, such data for African countries have been related to the census of population and some surveys.

Feasibility

The indicators should be chosen in such a way that they are relevant in the near future. The indicators that are too demanding in terms of statistics will serve only academic interest in most African countries. Feasibility may be evaluated in terms of the statistics that are or will become available in the foreseeable future.

Co-ordination and Structuring

It is desirable to base indicators on the co-ordinated and integrated basic statistics as far as possible. In order to avoid as many problems as possible, the indicators should use the same concepts and classification procedures as the basic statistics on which they are based.

Quantifiability

The indicators chosen should be quantifiable. Although this is not a particularly easy task, since traditions and customs affect the labour-force participation rates as well as the mobility of labour, to cite only two examples.

The indicators should not only be quantifiable, but also the scaling of the minimum level of human satisfaction is sometimes essential. For instance, it would be very useful if, together with the indicator of calories consumed, the minimum level of calorie requirements were also given. Similarly, some normative goal should be set for reducing inequalities in the country and be presented alongside the indicator of income distribution.

STATISTICAL TECHNIQUES FOR CONSTRUCTION OF SOCIOECONOMIC INDICATORS

The rearrangement of statistics and the adoption of suitable techniques for converting basic statistics into indicators, except in the few cases where basic statistics themselves serve as an indicator, form the subject matter of this section. The methods used in constructing the indicators vary from simple aggregations and percentages to synthetic index numbers, depending on the availability of data and the purposes for which the indicators are designed.

Simple Social and Economic Indicators

Simple social and economic indicators consist of simple aggregations, ratios and proportions, mean, median, mode, frequency distributions, skewness and kurtosis, and other simple measures. Indicators relate the numbers in a given state or experiencing a given event to the total population. It can be calculated as a ratio or as a rate; that is, the total number of subjects experiencing a particular activity is divided by the total number in the universe (i.e. those experiencing plus those not experiencing). This is a ratio indicator. When an interval within which the indicator operates is introduced into the definition, then rates can be calculated; crude birth rates and crude death rates are examples of indicators based on rates.

Measures of dispersion and central tendency are used in constructing indicators relating to the distribution of income and consumption. For measuring the severity of a given event and comparing it over time – for example, for the distribution of income among different social groups or classes – one requires measures such as Gini or Pareto coefficients. Data are generally not available in African countries for the measurement of these coefficients. Indeed, whatever data are available may be too scanty to construct a reliable and meaningful indicator.

Synthetic Index Numbers

Index numbers are used to construct indicators relating to output, prices, wages, and so on. Index numbers can be weighted or unweighted. However, for them to be useful for comparative purposes over time, weighting of the index numbers is essential. Weights may either be for the current period or for the base period, depending upon the purpose for which the index numbers are required. In African countries, the non-monetised proportion of the output and consumption can be substantial, but the prices of the goods can only be measured indirectly. Of course, there are many other problems in constructing index numbers of various activities in Africa, but these will not be discussed here.

Life Expectancies

Life expectancies as an indicator are commonly used for estimating the average length of time a person in any country in question is expected to live. The value of the expectation of life at age 0 is of special interest. It is the average length of life or expectation of life at birth and is frequently used as a convenient summary of the mortality experience depicted in the population. But in Africa, infant and child mortality are extremely high; and since this measure is an overall measure of mortality, the contribution of this section of the population (infants and children) affects the level of this measure of mortality considerably.

Life expectancies can also be used for other aspects of welfare, such as the length and achievement of education, the frequency and severity of health disabilities, the length of working life and the growth of income. Life expectancies can be based on cross-sectional data and on time series data. In Africa, where the statistics are still at the preliminary stage of development, it is extremely difficult to obtain time series data for any event.

The Taxonomic Method[1]

The taxonomic method for constructing social and economic indicators, developed by a group of Polish mathematicians, can be used for comparing development and the pattern of development among countries and among regions. It is a basically classificatory technique and uses measures of dispersion and central tendency for obtaining its coefficients. The method is very useful for economic planning among a host of its other uses. However, the method is fairly advanced, and its discussion is beyond the scope of our present discussion.

SOCIOECONOMIC INDICATORS IN AFRICA

Socioeconomic indicators can be constructed either by using existing data with all the constraints, or by building a model and then collecting the required data. The first approach has the merit of being more practical and feasible when data availability is used as a criterion. But we may end up constructing indicators that do not serve the purpose of the study fully. On the other hand, we may design indicators purely on the basis of an objective function; but when one starts collecting the data, one is faced with the same problem of data collection in Africa. The ideal is a combination of the two approaches. This compromise approach has the advantage of constructing indicators that can fulfil our objectives, in that we can use whatever data are available and supplement these by collecting any additional information instead of just collecting a new set of data.

Similarly, judgement or discretion is required in deciding on the number of indicators. If the number of indicators is too large, we may face the problem of synthesising them, apart from the problem of data requirements. On the other hand, if the number of indicators is too small, the aggregated index will hide the desired details. A compromise has to be struck while deciding on the number of indicators. In addition, the chosen indicators should fulfil the conditions laid down earlier in this chapter. Bearing in mind these factors, we have selected and now discuss a number of indicators that appear to us to be reasonable for African countries.

Gross National Product (GNP)

The reliability of the GNP (a measure that has been used more often than understood) will be discussed. Two aspects will come under scrutiny: first, its reliability; and second, its validity as a measure of development

The GNP is an aggregate of total output of goods and services. It is one of those amorphous terms that people think that they understand but often find difficult to articulate and, more important, to measure. As an aggregate, it is made up of several components, most of which are difficult to measure in Africa.[2] It is therefore pointless to argue that an aggregate such as the GNP is reliable, when the components that make up this aggregate are not themselves reliable. It is also arguable how reliable those national accounts are that are not based on regional or geographical accounts.

There is also the problem due to the fact that the *per capita* income is a ratio. It is the total national income divided by the total population of a country. Even if the numerator is assumed to be reasonably accurate and reliable, what about the denominator (i.e. population)? Assume that we find that the country's population is underestimated by 10 per cent. Will this imply that the *per capita* income is overestimated by 11 per cent? That is,

$$\textit{per capita}\text{ income} = \frac{100GNP}{90P} = 1{\cdot}11\,\frac{GNP}{P}$$

where *GNP* is the gross national product and *P* is the total population.

Another aspect of the problem of validity is the distribution of the

GNP within a country. Mahbub ul Hag (1971) has raised the problem of validity of *per capita* income in measuring development when remarking:

> We were confidently told that if we take care of our GNP, poverty will take care of itself. We were often reminded to keep our eyes focused on a high GNP growth target, as it was the best guarantee for eliminating unemployment and of redistributing incomes later through fiscal means. Then what really happened? Where did the development process go astray?
>
> My feeling is that we conceived our task not as the eradication of the worst forms of poverty but as the pursuit of certain high levels of per capita income. We conceived ourselves that the latter is a necessary condition for the former but we did not in fact give much thought to the interconnection. We, development economists, persuaded the developing countries that life begins at $1,000 and thereby we did them no service. They chased elusive per capita income levels, they fussed about high growth rates in GNP, they constantly worried about 'how much was produced and how fast', they cared much less about 'what was produced and how it was distributed'.

To illustrate the point, consider the following example. Suppose that an economy consists of only ten people, nine of whom have no income at all, while the tenth receives 100 units of income. GNP will be 100 and *per capita* income 10. Now, suppose that everyone's income increases by 20 per cent, so that the GNP rises to 120 while *per capita* income grows to 12. For the nine individuals with no income before, still no income; the 20 per cent rise in the *per capita* income will provide little comfort. The one rich individual will still have all the income. In this case, GNP growth, instead of being an index of social welfare growth, merely measures the welfare growth of a single individual. The same analysis can be applied to the more realistic situation in which incomes are very unequally distributed, although not perfectly unequal as above.

The example given deals with GNP *per capita* in a single country. As so often happens, GNP *per capita* is also often used for comparing living standards in different countries. But the exercise has statistical usefulness only if the following limitations are borne in mind. First, the national average should be accompanied by some data on the distribution of income, unless we can assume that the distribution of income in the two countries is approximately the same. There is a great deal of difference in terms of real living standards between a country with a *per capita* income of, say, US $400 per annum where income distribution is approximately equal, and one where 20 per cent of the population have an income per head of US $1,500 and 80 per cent an income of US $125. Second, in considering living standards, changes in the distribution of income over time may have as great an effect as changes in the gross product.

The *per capita* income, although a crude measure of development, because of these limitations, is still used as an indicator of development; so, let us examine how it behaves under the impact of population growth. Let us take two countries, A and B; let us assume that total income doubles itself after every twenty years. Assume further that the population in the case of country A doubles after twenty years, but that in country B the population doubles every fifty years. Let both countries start with a *per capita* income of US $100.00.

At the end of twenty years, *per capita* income will still be US $100 in country A and US $152 in country B. After forty years, country A will still be US $100 and country B will be US $230. The conclusion is clear, and it is: all other things being equal, if total income is growing at the same rate in two countries A and B, *per capita* income will be growing more rapidly in the country with the slower rate of population growth. All this, of course, assumes that the income distribution is the same in the two countries.

Bondestam (1973) has made the point that:

> Some of the weaknesses of G.N.P. and G.N.P. per capita, which have been elaborated on, imply a definite risk of biased descriptions of countries' economies when utilising these standards as measures of wealth. This bias is two-fold: first, as we have seen from the Ethiopian example above, the difficulties in estimating the production in the various sectors of the economy, particularly the predominant peasant subsistence agriculture, and in objectively valuing the unit-price of the product may lead to wrong conclusions when it concerns the average level of economic wealth of the respective country.
>
> This bias in the conclusions is further accentuated by difficulty, or impossibility, to weigh incomes against expenditures, of which the latter vary remarkably from country to country. Second, as was shown in the Kenyan example earlier average figures for the economic status of a whole population not only hide the uneven distribution of wealth, but may even give an entirely erroneous and false impression of the economic standard of living of the masses. This criticism of G.N.P. and G.N.P. per capita gives rise to an immediate question: Is there any alternative standard? Which are the other and better indicators of wealth with which G.N.P. could be replaced?
>
> The question is political – simply because the underlying causes of under-development are political. Moreover, in order to answer

it, we first have to define concepts like under-development and wealth, an elaboration of such magnitude that it definitely lies outside the scope of this paper. Despite these reservations an attempt will be made, where some alternatives to G.N.P. are suggested. For this purpose the population has to be divided into a number of groups, either to their economic activity or status (subsistence farmers, commercial farmers, pastoralists, wage-labourers, owners of the means of production, merchants, officials in government service, etc.).

Seers (1972) has agreed to the need for an alternative definition to development and suggested how this should be measured. He has commented as follows:

> Development means creating the conditions for the realisation of human personality. Its evaluation must therefore take into account three linked economic criteria: (a) poverty, (b) unemployment, (c) inequality. G.N.P. can grow rapidly without any improvement on these criteria; so development must be measured directly.

While there is a general agreement on using other criteria for measuring development indicators, we are not unanimous on the selection of these indicators. Bondestam (1973) has given some alternative measures and remarked that:

> Some of the alternative measures of development, which could be used instead of the overall G.N.P. per capita, are: division of economic wealth among the various economic groups, measured in, for example, purchasing power and related to a defined minimum standard of living; the proportion of the population, which is under constant threat of famine; the economic possibilities of the various economic groups to fulfill the most immediate daily needs, like housing, sufficient calorie and protein intake, and clothing; the proportion of the population within the monetary sector of the economy; distribution of economic growth among the various groups; economic dependency between the various economic groups, between rural and urban areas, between the actual country and the industrialized countries, etc.; trends in concentration or deconcentration of capital, land and wealth; type of industry (light or heavy), rate of investments, and rate of industrialization, measured in production as well as employment; security against confiscation of land, eviction from land, loss of job, etc.; access to medical and health services, education, community development services, technical assistance in agriculture, etc.

These are a few of the possible alternatives to G.N.P. as measures of degree of under-development. Although they are not necessarily the best, they are at least better than G.N.P. since they take care of the distribution of wealth and some of them to a certain extent entail the causes of under-development, which the G.N.P. does not do.

We therefore give a description of other developmental variables used as indicators apart from the GNP *per capita*, which we have already discussed.

Population

African populations are growing very fast. The high population growth rates are of the order of 2·5 per cent per annum.[3] There is no urgent desire to introduce modern family-planning methods and population-planning in African development plans. Nevertheless, the study of the growth of African populations, their birth rates and death rates, is important apart from the academic interest, for planning purposes: for example, provision of facilities that the population may require. It is also important for estimating the manpower requirements in the country. Indicators such as age and sex ratios, dependency ratios, and the gross and net reproduction rates are useful in demographic analysis. Other indicators, such as those measuring mortality, reflect health conditions in any given country.

An important aspect of the population-planning in African countries, which requires urgent attention, is internal and international migration. Rural – urban migration and spatial distribution, and so on, require urgent attention. Indicators such as rates of urbanisation, in-migration rates, out-migration rates and net migration rates are therefore required for planning purposes.

Health and Nutrition

In planning for health, one is concerned with the medical facilities available to the population. Indicators of health therefore measure both the health conditions and the improvement in health conditions of the population over time. This can be indicated by the number of physicians and surgeons as well as hospital beds available per 100,000 of population. Indicators dealing with the nutritional standards of the population are also required; for example, an indicator for the quantity of food taken is the daily calorie consumption *per capita*, and an indicator for the quality of food taken is the daily protein consumption *per capita*.

Housing and Environment

Housing is needed for shelter, and so on, and its improvement forms the social and economic objective of any planning process. Indicators of housing conditions that can be calculated are: the percentage of population living in permanent (cement) houses; percentage of population living in the temporary structures; percentage of population without any house and facilities such as electricity, water, toilets and bathrooms; average number of persons per room, and so on.

Education and Culture

The importance of education can hardly be overemphasised, and so is the need for indicators on education required for any planning purposes. Enrolment ratios and gross enrolment ratios can serve as useful indicators. The gross enrolment ratio relates total enrolment at each level to the number of persons in corresponding age groups.

Females as a percentage of total primary enrolment are another indicator. As an indicator of female education, this enrolment ratio is quite useful. This ratio serves as a useful indicator, particularly when it is calculated at the second or at the third level of the educational ladder, because primary education is more or less compulsory in most African countries, although female education is still unpopular. For example, although education is compulsory in Nigeria, female enrolment ratio at the primary level is low, particularly in northern Nigeria.

Other indicators required in Africa are the student – teacher ratio, input – output ratio and graduation rates. Adult education in African countries is becoming popular; therefore, an indicator showing the adult education rate, covering the population of 15 years or above, would also be required for educational planning.

Cultural indicators are also required; for example, the ratio of leisure time utilised for cultural activities, the number of radio and television receivers per 100,000 of population, and the availability of public libraries, playgrounds, and so on can be used as indicators of culture in a given country.

Employment, Working Conditions and Social Security

The relevant indicators that reflect the employment aspects of planning are: percentage of population of working age who are economically active; percentage of economically active population unemployed; percentage of the unemployed who have completed secondary-level education or above; percentage of women in non-agricultural employment; percentage of population self-employed, and so on.

Indicators of conditions of work that can be used are the number of hours of work, number of industrial disputes, real wages, industrial accidents and paid holidays. Social security schemes are in their infancy in Africa, and data are scarce. Crude indicators of social security can therefore be taken as the percentage of the economically active population covered by social security benefits and proportion of eligible persons in receipt of such benefits, as well as benefits as a proportion of the available income.

Social Defence and Welfare

Crime rates and juvenile delinquency rates are among the principal indicators of crime. These are usually compiled from police records. Government expenditure per 1,000 of population on the police and the law enforcement agencies can also serve as an indicator of the government effort to reduce crime and increase social defence.

Public welfare, as its name suggests, can be taken as public help to victims of crime or floods, handicapped persons, orphans, destitutes, and so on. It covers a wide range. It is therefore very difficult to construct a common indicator of social welfare. The best that can be done is to use government and private institutional expenditure on social services per 1,000 of population.

Income and Wealth

The distribution of personal income and personal wealth is an emotive subject on which views of social justice, income inequalities and growth all have a proper place. A problem limiting discussion has always been the difficulty of assembling acceptable statistical material. Income distributions have proved more tractable and have been subjected to much analysis, but the possibly more significant aggregate, personal wealth, has always proved more difficult to estimate.

Indicators of wealth are therefore very difficult to compile. There is an almost complete absence of data on all components of wealth, except for a few items. Data on the financial components of wealth are easy to get. These can be obtained from the statements of accounts of banks – both commercial and central. But there is a complete statistical vacuum on data concerning the other components of wealth in African countries. However, time and again figures produced for one purpose have to be applied to another; for example, estimates of the distribution of wealth can be obtained from tax returns and used as indicators of wealth, but this can only be regarded as a stop gap measure and a very approximate estimate. The amount of sheer numerical analysis, the detailed sifting of base data, the consequences of differing assumptions and the reconciliation of alternative estimating methods are enormous.

The United Nations Research Institute of Social Development (1970) has brought out a list of social and economic indicators. The list is by no means exhaustive, but rather suggestive of possible items that can be used for measuring development. Some of these are: expectation of life at birth; percentage of population in localities of 20,000 and over; consumption of animal protein *per capita* per day; combined primary and secondary enrolment; vocational enrolment ratio; average number of persons per room; newspaper circulation per 1,000 of population; percentage of economically active population employed in electricity, gas, water, and so on; agricultural production per male agricultural worker; percentage of adult male labour force in agriculture; electricity consumption (kilowatts) *per capita*; steel consumption (kilograms) *per capita*; energy consumption (kilograms of coal) *per capita*; percentage of gross domestic product (GDP) derived from manufacturing; external trade *per capita* (in US dollars); and percentage of salaried and wage-earners to total economically-active population.

The availability of indicators on these items in African countries has been given in the *Survey of Economic Conditions in Africa, 1971 – 72*, published by the Economic Commission for Africa (ECA), and readers are advised to consult this study for their respective countries. We shall, however, attempt to give some general indications of what we see as inadequacies and shortcomings in the data on socioeconomic indicators. Reviewing the deficiencies, it is possible to break down the problem into specific areas. There is mis-specification as with national income accounts and inaccuracy as with crime figures, and a host of other social and economic statistics and levels of disaggregation that are poor. There is, for example, the need for better and disaggregative measures of social and economic health of each country, such as the proportion of the population below poverty level, the distribution of income and wealth among various income and ethnic groups and regions, the degree and pattern of labour utilisation, and the nature and extent of malnutrition.

The analysis in the *Survey of Economic Conditions in Africa* clearly shows the lack of statistical material to construct socioeconomic indicators in African countries. Part of the answer must lie in the provision of more figures. More data on the transactions within the economy would therefore be welcome. However, in view of the high cost of statistical works the need is for a greater use of administrative data and a greater effort on the analysis and interpretation of existing data.

As a matter of urgency, more information is clearly required on vital statistics, employment, unemployment, income distribution, migration and those data relating to the household. It is necessary, however, to take a long term view of official requirements for new statistics. The aim is not necessarily to include a mass of new information or data, but to ensure that the design of the reporting system leaves open the opportunity to bring in whatever kinds of information may be needed in the foreseeable future.

Two factors, however, have a bearing on our judgement of what statistics to collect: first, technology can almost certainly be counted on to cut down costs of producing and collecting figures (e.g. processing); and second, data compilation is only the start of a whole chain of processes, which include interpretation of the data collected, statistical analysis and research, which are very dependent on human beings, even if they make use of machines. Notwithstanding these problems, every African country is trying to develop, within its limitations, the collection of better data. A further problem, however, is how to reconcile the needs for international comparability, and hence conformity with the standards laid down by the international agencies, with the specific domestic needs of an individual country.

CONCLUDING REMARKS

The main purpose of socioeconomic indicators is to provide essential background data needed by those concerned with policy formulation and monitoring. They aim to show where society has to go and how it is changing; and they try, wherever possible, to highlight interactions and changes in relationships.

We have made a number of important points about the selection of socioeconomic indicators. The first most important point is that the selection of indicators should be related to their purpose and, in the context of our discussion, to their usefulness in development planning: that is, for analysing the situation and for defining and evaluating policies and programmes. The second important point is that the development of indicators should be closely related to the needs of planners on the one hand and the supply of basic information on the other.

However, the construction of socioeconomic indicators to measure development in a comprehensive way at national and regional levels is at the experimental stage. Much more experimentation is needed. An area where research must be focused is the problem of disaggregation, both spatially and by socioeconomic groups. However, priorities will have to be decided, depending on the level of statistical development, the administrative structure, the kind of planning, and so on.

Of course, a major change in development policy would imply a corresponding change in the output and the organisation of an official statistical office. This raises general issues to all African countries. What is involved therefore is not just a switch of emphasis in the collection of data,[4] but a change in the conceptual framework. This, of course, raises critical issues, some of which we have discussed in the previous chapters. It is, however, to the improvement of data collection procedures for measuring development policies and for the construction of sensitive socioeconomic indicators that we now turn our attention in the closing remarks.

In the field of data collection, African countries should institute national household inquiries on a regular basis. Demographic-data collection programmes can also be established, and a good deal of work can be done on the collection of data on local food production and consumption. The collection of data will need to be supplemented by research in greater depth on the patterns of household consumption and the use of labour, output, incomes and marketing in various parts of the country, if economic policy is to focus properly on the relief of poverty.

There is no doubt that a national household survey offers to an African country a number of fundamental advantages in the collection and analysis of its social and economic statistics. Once an appropriate sampling frame has been designed and a satisfactory field organisation created – basic steps in establishing a capability – a programme consisting of ongoing survey rounds may be designed, and universal but flexible coverage of the population is facilitated.

Data can be gathered on the entire population of a country, including the rural population and the traditional sector, or, if required, on special population categories that may be of particular interest for use in policy or programme analysis. Such population categories can be selected according to any geographical, income, employment, sex or other characteristics considered important in the national or local circumstances.

A continuing multiround survey or programme of surveys may cover a wide range of subjects and, if properly designed, can accumulate invaluable measurement experience, on the one hand, and trend or updated information, on the other. Such continuing survey-taking activities also lend themselves to step-by-step innovation and experiment and to the detailed investigation of such complex topics as income and employment, nutrition, literacy, health and other elements affecting the levels of living. As the cumulative process advances, the marginal unit cost of additional and improved data output is expected to drop considerably, depending, of course, on the nature of the successive rounds or surveys. Likewise, as experience is gained in running a permanent field organisation and data-processing activity, timeliness can be enhanced, to the point where key series can be made available soon after field work and where the policy-maker's urgent needs for data on important topics can be met without unnecessary delay.

However, this is not all. The household plays a key role in development in Africa. Not only is it in the household that the impact of development on living conditions must be measured, but it is there that much productive activity is organised, including most of the agricultural sector and the so-called 'traditional' sector. A national household-survey capability is the only systematic way to investigate the household as a complex economic and social unit and to measure its role in national activities.

With the ability to conduct properly organised and continuing surveys, statistical services would be in a position to supply integrated data on employment, other productive activity, income, consumption and expenditure, together with related social and demographic statistics required for the assessment of problems and issues in such areas as rural development, mass poverty, literacy, population programmes and maternal and child health. It further has to be borne in mind that the permanent field organisation that will be established or upgraded will also contribute to the improvement of agricultural, industrial, employment, price and other statistics. Thus, a continuing survey organisation provides a basis for securing up-to-date and integrated information for policy formulation and for programme-planning and evaluation. Moreover, it provides a vehicle for experimentation with concepts, definitions, questionnaires, and so on within the context of the diverse needs and circumstances of individual countries.

NOTES: CHAPTER 13

1 For the application of the taxonomic method to live data, see Ewusi (1975b).
2 See Chapter 12 for a description of the problems of measuring GNP in Africa.
3 See, for example, Population Reference Bureau (1975).
4 There are certainly some severe deficiencies in the statistical description of the economies of African countries, and possibly some distortions also. This is because of the bias towards data on things that can easily be measured. The outcome of these deficiencies has been very unsatisfactory forecasting of the economy.

REFERENCES AND FURTHER READING

Bondestam, L. (1973), *Some Notes on African Statistics: Collection, Reliability and Interpretation*, Research Report No. 18 (Uppsala: Scandinavian Institute of African Studies).

Clift, F. C. (1975), 'New perceptions of the problem of development', paper presented at the conference on Statistical Policy in Less Developed Countries, Institute of Development Studies, University of Sussex, Brighton (12 – 16 May) IDS No. 114.

Dasgupta, B. and Seers, D. (1975) (eds), *Statistical Policy in Less Developed Countries*, IDS No. 114 (Brighton: Institute of Development Studies, University of Sussex).

Economic Commission for Africa (undated), *Survey of Economic Conditions in Africa, 1971 – 72* (Addis Ababa), pt 2.

Enabor (1976), 'Wood consumption requirements in Nigeria: a reassessment', *Economic and Social Studies*, vol. 18, no. 1, pp. 121 – 45.

Ewusi, K. (1975a), 'Report of the workshop on socio-economic indicators', paper presented at the conference on Statistical Policy in Less Developed Countries, Institute of Development Studies, University of Sussex, Brighton (12 – 16 May), IDS No. 114.

Ewusi, K. (1975b), 'Analysis of socio-economic variables as indicators of development', paper presented at the conference on Statistical Policy in Less Developed Countries, Institute of Development Studies, University of Sussex, Brighton (12 – 16 May), IDS No. 114.

Fapohunda, O. (1975), 'Report of the discussion on the workshop report on socio-economic indicators in the plenary session', report presented at the conference on Statistical Policy in Less Developed Countries, Institute of Development Studies, University of Sussex, Brighton (12 – 16 May) IDS No. 114.

Frankel, S. H. (1953), *The Economic Impact on Under-developed Societies* (Oxford: Blackwell).

International Labour Organisation (1972), 'Statistics and research', in *Employment, Incomes and Equality* (Geneva), ch. 31.

Killick, T. (1978), *Development Economics in Action: A Study of Economic Policies in Ghana* (London: Heinemann), ch. 4, table 4.3.

Mahbub ul Hag (1971), 'Employment in the 1970s: a new perspective', *International Development Review*, no. 4.

Population Reference Bureau (1975), *World Population Data Sheet* (Washington, DC, annual).

Seers, D. (1972), 'What are we trying to measure?', *Journal of Development Studies* (April).

Stone, R. (1975), 'Social statistics and social policy', paper presented at the conference on Statistical Policy in Less Developed Countries, Institute of Development Studies, University of Sussex, Brighton (12 – 16 May), IDS No. 114.

Todaro, M. (1977), 'Development policy and population growth: a framework for planners', *Population and Development Review*, vol. 3, nos 1 and 2, pp. 23 – 43.

United Nations (1970), *Contents and Measurements of Socio-Economic Development*, Report No. 70.10 (Geneva: Research Institute of Social Development).

United Nations (1976), *Social and Demographic Indicators: Draft Guidelines on Social Indicators*, E/CN.3/488 (New York: UN Economic and Social Council).

United Nations (1978), 'UN household survey capability: a progress report', *International Statistical Review*, vol. 46, no. 3, pp. 307 – 11.

Index

Notes Page numbers in **bold** type are references to major subjects. Page numbers in *italic* type refer to tables, figures, etc.

For Product Safety Concerns and Information please contact our EU
representative GPSR@taylorandfrancis.com
Taylor & Francis Verlag GmbH, Kaufingerstraße 24, 80331 München, Germany

www.ingramcontent.com/pod-product-compliance
Ingram Content Group UK Ltd.
Pitfield, Milton Keynes, MK11 3LW, UK
UKHW022155190926
475702UK00003B/43

* 9 7 8 1 0 3 2 7 4 7 0 1 9 *